T0223314

G. Ottaviani (Ed.)
Financial Risk in Insurance

Springer
Berlin
Heidelberg
New York
Barcelona
Hong Kong
London
Milan
Paris
Singapore
Tokyo

G. Ottaviani (Ed.)

Financial Risk in Insurance

With 20 Figures

 Springer

G. Ottaviani (Ed.) †
Istituto Italiano degli Attuari
Via del Corea 3
00186 Rome, Italy

Authors:

Phelim Boyle

Director of the Centre for Advanced Studies in Finance
University of Waterloo, Waterloo, Ontario, Canada N2L 3G

Hans Bühlmann

Department of Mathematics, ETH Zentrum, 8092 Zürich, Switzerland

Massimo De Felice

Dipartimento di Scienze Attuariali e Finanziarie, Facoltà di Statistica
Università di Roma, "La Sapienza", Via Nomentana, 41, 00161 Roma, Italy

Franco Moriconi

Istituto di Matematica Generale e Finanziaria
Facoltà di Economia
Università di Perugia, Via Pascoli, 1, 06100 Perugia, Italy

Flavio Pressacco

Dipartimento di Finanza della Impresa e dei Mercati Finanziari
Facoltà di Economia
Università di Udine, Via Tomadini, 30A, 33100 Udine, Italy

Mathematics Subject Classification (1991): 60-00, 62-00, 62P05

Cataloging-in-Publication Data applied for
Die Deutsche Bibliothek – CIP-Einheitsaufnahme

Financial risk in insurance / G. Ottaviani (ed.). – Berlin; Heidelberg; New York; Barcelona; Hong Kong; London;
Milan; Paris; Singapore; Tokyo: Springer, 2000
ISBN 3-540-66143-3

ISBN 3-540-66143-3 Springer-Verlag Berlin Heidelberg New York (Softcover edition)
ISBN 3-540-57054-3 Springer-Verlag Berlin Heidelberg New York (Hardcover edition 1995)

Typesetting: Protago · TEX · Production, Berlin
Cover design: *design & production* GmbH, Heidelberg

Printed on acid-free paper   SPIN 10723993   41/3143 – 5 4 3 2 1 0

Preface to the Soft-Cover Edition

When "Financial Risk in Insurance" appeared in 1995, we would not have imagined that this text would find such a wide readership. After all actuarial colleagues had received the text automatically through their subscription to the 1993 AFIR colloquium in Rome. So the demand must have come from outside of our own professional circles, we believe from researchers and practitioners in finance. Both in 1996 and 1997 further copies needed to be printed.

We therefore applaud the initiative by Springer to make this text available in the form of a soft-cover edition. We hope that this new edition will further contribute to the very fruitful dialogue between actuaries and professionals in finance and will be helpful in the cultural thought process bringing the world of banking and insurance closer to each other.

Zürich, 1 June, 1999

In the name of the authors

Hans Bühlmann

Preface

The Istituto Nazionale delle Assicurazioni (INA), a leading company on the Italian life insurance market for over eighty years, takes special pleasure in sponsoring this scientific volume meant for the large international community of those concerned with insurance and finance.

Our involvement in this initiative is directly connected with the awareness that the domain of insurance, in particular with respect to the management of long-term insurance savings, is changing. This enlargement, emphatically noticeable in the area of life insurance and pension funding, is extending to cover also the "interest rate risk".

The path for this evolution is evidently set. It clearly also offers new opportunities for the professional development of those working in insurance, in particular the actuaries. There is no longer any doubt that life insurance assumes heavily the role of a system which provides adequate guarantees against investment risks. To the matching techniques between assets (investments) and liabilities (technical reserves) – belonging to the core of traditional actuarial activities world-wide – the new techniques of immunisation of investments and optimisation of results, e.g. by revoking to instruments of modern finance (e.g. options, swaps, futures etc.), need to be added.

The studies included in this volume are meant in such a perspective. They represent useful educational material for insurance companies and particularly for their actuaries. The latter ones will – e.g. in the context of the single European insurance market – be called upon to assume new significant functions especially with regard to the investment policy and the financial solvency testing of the companies. Incidentally and with great pleasure, I would like to point out that the authors of two of the lectures included in this publication were, in the past, awarded the distinguished International Prize INA – Accademia dei Lincei.

I trust you will allow me to interpret the above as the best possible confirmation of the foresight by INA's endeavours to support such scientific activities as are bound to contribute to the development of insurance, particularly also in life insurance.

Lorenzo Pallesi
President
Istituto Nazionale delle Assicurazioni - INA

Introduction

When AFIR (the financial section of the International Actuarial Association, IAA) assigned to the Italian Actuaries Institute the task of organizing the 3rd International AFIR Colloquium, the Italian actuarial community was highly pleased with the prestige of the occasion, conscious of the pleasure given by such hospitality.

In my personal satisfaction there was also a considerable historical and scientific component. Working on the scientific project of this Colloquium induced me to reflect on the reasons that had led to the establishment, the activity and the Colloquia of AFIR. It came naturally for me to open up a historical perspective, to reason, compare and assess the features of today's cultural situation with reference to the basic points of Italy's actuarial tradition, which may be read, in brief, in the "goals" of the School of Statistical and Actuarial Sciences, founded in the then Royal University of Rome almost sixty-five years ago.

AFIR came into being to characterize an "actuarial approach for financial risks" and marks the evolution of the actuarial species towards the "actuaries of the third kind". The representatives of this new typology of researcher and professional (going back to the identikit proposed by Hans Bühlmann in the now famous editorial published in the Astin Bulletin) are "mathematical experts who unfold their skills on the investment side of insurance or banking" capable of weighing up financial risks, conscious designers of synthetic or artificial products aimed at risk management (derived products, futures, options, etc.).

"The probability background already essential to the Actuary of the Second Kind must be substantially enlarged for the professionals of the Third Kind. Such notions from the theory of stochastic processes as stochastic integration, Ito formula, Black Scholes formula must be at hand in the latters' tool kit"[1] .

There emerges the figure of an actuary characterized by a knowledge of mathematical methods, of probability and of the theory of stochastic processes, of the statistics of processes, with a sound economic culture and a clear awareness of the mechanisms of trade, with a constructive attention towards business management problems.

If one does not consider the particular connotations induced by the current institutional context (specificity of financial contracts, market structure, etc.), which might distort one's judgment, one sees clearly that the Actuary of the third kind is in line with the best actuarial tradition. I was glad to find in this new figure of researcher and professional the signs of an inspiration that was also that of the Italian School.

In the January 1931 issue of the Giornale dell'Istituto Italiano degli Attuari, Professor Guido Castelnuovo set two goals for the School of Statistical and Ac-

---

[1] Bühlmann, H., Actuaries of the Third Kind?, Astin Bulletin, 17,2, November 1987.

tuarial Sciences, recently established in Rome, of which he had been appointed Director.

A scientific goal was set: to promote the study of Probability and of its applications, to promote in general the applications of mathematics to Statistics and to Political Economy. The professional objective was emphasized: "to form the actuary (in the broadest sense of the term), who should possess a solid mathematical and economic background".

To specify the effects of this inspiration on the formative project, it could be useful to recall that at that time the basic subjects taught by the School were: Probability, Actuarial Mathematics and Mathematical Statistics, Statistics and Political Economy (taught by the illustrious professors Guido Castelnuovo, Francesco Paolo Cantelli, Corrado Gini and Rodolfo Benini). It should be added that "apart from these university subjects, the students of the School usually attend the lectures on insurance culture which, by a discerning initiative of the INA, National Insurance Institute, are held periodically at the Institute's premises, and at which topical problems regarding Statistics, Political Economy or Insurance are discussed" [2].

This reference to the origins is important if it can give strength to the actions under way, and to me it has been a proficuous reference in starting the scientific project of this Colloquium.

This book contains the invited contributions presented at the 3rd International AFIR Colloquium, held in Rome in 1993.

The scientific programme of the Colloquium was aimed at encouraging research on the theoretical bases of actuarial sciences, with reference to the standpoint of the theory of finance and of corporate finance, together with the methods of mathematics, of probability and of the theory of stochastic processes. In the spirit of actuarial tradition, it was hoped that there would be a link between the theoretical approach and the operative problems of financial markets and institutions, and insurance ones in particular.

Three main themes were chosen as the basis on which the Colloquium was planned: analysis of financial markets and financial products, models and techniques for financial risk management, and the effects of the modern theory of finance on financial institutions.

The "invited contributions" were requested to provide a perspective of the overall research themes, able to contribute towards defining the cultural climate of the Colloquium, to be a boost to further study by young actuaries and points of comparison (especially in terms of methodological choices) for the actuarial profession.

With reference to the first theme, reports were presented on "Life insurance with stochastic interest rates", by Hans Bühlmann, and on a scheme for "Analyzing default-free bond markets by diffusion models", by Franco Moriconi. On the second theme, Phelim P. Boyle tackled the problems of defining and measuring "Risk-based capital for financial institutions" and Massimo De Felice

---

[2] Castelnuovo, G., La Scuola di Scienze Statistiche e Attuariali della R. Università di Roma, Giornale dell'Istituto Italiano degli Attuari, 2, 1, 1931.

proposed "An actuarial perspective on asset-liability management" within the methodological framework of the "Immunization theory". For the third theme, Flavio Pressacco proposed an analysis dedicated to "Financial risk, financial intermediaries and game theory".

Originally six "invited contributions" had been programmed, two for each of the main themes. Due to "reasons of state", Professor Joseph Stiglitz (appointed to a prestigious governmental office) was unable to participate in the Colloquium. In the idea of proposing the third theme and in the scientific characterization we gave it, his contribution as an expert on markets and institutions, while not submitted, is nevertheless potentially present.

It has given me pleasure to note that in all the contributions, in addition to clear doctrine there is appropriate reference to facts and data, and a sensitivity towards the construction, application and use of models. I am confident therefore that also for the "actuarial approach to financial risks" it will be possible to exploit for that same natural connection between formal techniques and economic meanings that has marked the success of stochastic calculus in the option pricing theory, if first and foremost the capacity spreads of not confusing rigour in a mathematical sense with rigour in an economic sense.

Rome, December 1993

Giuseppe Ottaviani
Chairman of the Istituto Italiano degli Attuari
Chairman of the 3rd International AFIR Colloquium

Contents

Life Insurance with Stochastic Interest Rates

Hans Bühlmann

Mathematik, ETH, 8092 Zürich, Switzerland

1 Classical Mathematics of Life Contingencies

1.1 The Individual Policy

The basic building stones of the model are the individual policies. Each policy defines a stream of stochastic cash payments in the time points $0, 1, \ldots, n$.

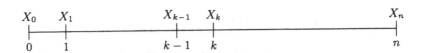

We summarize the cash stream in the *cash payment vector*

$$X = (X_0, X_1, \ldots, X_n)$$

and interpret the components of X as follows

$$X_k = \text{Benefits in } (k-1, k] - \text{Premiums in } k + (\text{Costs in } (k-1, k])$$

Observe: $X_k > 0$ means a positive payment made by the insurance carrier.

Remark.

- I have used the Swiss convention which assumes cash payments (and cost payments) *at the end* of the interval $(k-1, k]$. Obviously one could work with other conventions (e.g. middle of the interval)
- Depending on whether or not *costs* are included or excluded we have a gross or a net model.

1.2 Valuation at Time Zero ($t = 0$)

With the standard discount factor $v = \frac{1}{1+i}$ one defines the expected discounted present value of the cash payment stream

$$\text{Val}[X] = \text{E}\left[\sum_{k=0}^{n} v^k X_k\right]$$

The Equivalence Principle can then be formulated

$$\text{Val}[X] = 0\,,$$

which especially allows the calculation of *net* and *gross premiums*.

1.3 Valuation at Time $t > 0$

With each policy we associate the filtration

$$\mathcal{F}_0 \subset \mathcal{F}_1 \subset \ldots \subset \mathcal{F}_{t-1} \subset \mathcal{F}_t \subset \ldots \subset \mathcal{F}_n$$

where $\mathcal{F}_t$ represents the "information" available at time t; mathematically we speak of an increasing sequence of σ-algebras. The idea that "X_k is known at time k" is translated into this mathematical language as "X_k is $\mathcal{F}_k$-measurable".

$$\text{Val}[X \mid \mathcal{F}_t] = \underbrace{\text{E}\left[\sum_{k=0}^{n} v^{k-t} X_k \mid \mathcal{F}_t\right]}_{\text{conditional expectation}}$$

$$= \sum_{k=0}^{t} r^{t-k} X_k + \text{E}\left[\sum_{k=t+1}^{n} v^{k-t} X_k \mid \mathcal{F}_t\right]$$

$$= \text{A}\left[X \mid \mathcal{F}_t\right] + \text{R}\left[X \mid \mathcal{F}_t\right]$$

with $r = v^{-1}$. The interpretation is

- $A[X \mid \mathcal{F}_t]$ are the accumulated payments made by the insurance carrier
- $R[X \mid \mathcal{F}_t]$ is the prospective reserve

One can show that

$$-A[X \mid \mathcal{F}_t] = R[X \mid \mathcal{F}_t]$$

"on average" (i.e. by taking expectations) i.e. payments received equal the prospective reserve "on average".

Deviations from the expected quantities are defined as losses, more precisely:

- deviations due to X_k's $\Rightarrow$ technical loss (gain)
- deviations due to realized interest being not equal to r $\Rightarrow$ financial loss (gain)

One must observe that **technical losses** are defined *inside* the mathematical model, whereas **financial losses** occur *outside* the model. This is one of the reasons why we need to incorporate *stochastic interest rates* into the model of Life Contingencies. In the following I want to show how to do this; in particular I want to answer the following two questions *within* a model of life contingencies based on stochastic interest rates:

1. How to separate technical and financial loss?
2. Is it reasonable to keep the reserving basis (technical interest rate) independent of experience made?

1.4 Critique of the Classical Model

a) Critique from Actuarial Considerations. Why is the technical interest value i assumed to be the same for all years? Why should a cautious actuary use the same discount factor v for discounting

i) from year one back to year zero
ii) from year thirty back to year twentynine?

obviously the operation ii) is accompanied by a *much greater uncertainty* than operation i).

b) Observation Made on the Financial Markets. On financial markets interest rates depend clearly on the *term* of the contract. (term structure of interest rates) Economists think of the following situations:

i)   ii)

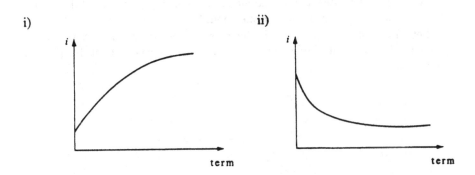

and call them

i) "standard term structure"
ii) "inverse term structure"

For us as actuaries ii) seems the more natural one as it is expressing the fact that uncertainty in discounting is increasing with time.

Obviously one can also have

iii) "mixed term structure"

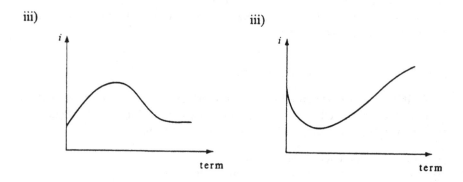

iii)   iii)

It is remarkable that both critiques a) and b) *disappear* in a (reasonable) model which incorporates stochastic interest rates.

2 Approach via Stochastic Interest Rates

It is worthwhile to understand that one has to be careful, when working with stochastic interest rates. new phenomena, not known in a world of deterministic interest rates, appear, which one should clearly recognize.

Let me illustrate my point in the most simple situation: "What is the present value of 1 unit of money to be received in one year?" Call this present value x.

```
money        x                        1
             ├──────────────────────┤
time         0                        1
```

Solution A:

- If the interest i would be known

$$x = \frac{1}{1+i} \, .$$

- If interest i is stochastic then take expectation, hence

$$x = \mathrm{E}\left[\frac{1}{1+i}\right] .$$

Solution B:

- If the interest i would be known, x would grow into $x(1+i)$ in one year.
- If interest i is stochastic then it grows into $x\,\mathrm{E}[1+i]$ "on average" which should be equal to 1. Hence

$$x = \frac{1}{\mathrm{E}[1+i]} \, .$$

Who is right? The person who choses A or B?

Remark: Do not think that the fact, that we have used *expectation* to construct the certainty-equivalent, is responsible for the paradoxon. One can construct it with all reasonable functionals of the distribution of i.

The lesson to be learned is that one can not take expectations over both the discount factor *and* the capitalization factor. I shall come back to the paradoxon. For the moment it does however strongly imply that we need to construct a theory of stochastic interest starting *from first principles*; otherwise we might fall in the trap of similar contradictions as the one just explained.

2.1 Valuation at Time Zero (Q[X])

As I have used Val[X] for the valuation with fixed constant interest rates I use now the symbol Q[X] for the valuation in the presence of stochastic interest. We *postulate* that Q[X] should be a positive *continuous linear functional*. This postulate can be economically justified either from the "*non arbitrage principle*" or from *equilibrium theory*. More mathematically:

Assume $X = (X_0, X_1, \ldots, X_n) \in L_{n+1}^2$ (which is a Hilbert space) i.e. assume

$$\mathrm{E}\left[\sum_{k=0}^{n} X_k^2\right] < \infty \tag{1}$$

In this Hilbert space the scalar product is defined as

$$(X, Y) = \mathrm{E}\left[\sum_{k=0}^{n} X_k Y_k\right] \tag{2}$$

As in all Hilbert spaces identify almost equal vectors i.e.

$$X \cong Y \quad \text{if} \quad X = Y \quad \text{almost surely} \tag{3}$$

As any continuous linear functional can be written as a scalar product we must have

Representation Theorem. There is a $\varphi = (\varphi_0, \varphi_2, \ldots, \varphi_n) \in L^2_{n+1}$ such that

$$Q[X] = \mathrm{E}\left[\sum_{k=0}^{n} \varphi_k X_k\right] \tag{4}$$

and as $Q[X]$ is supposed to be a positive functional we must have $\varphi_k > 0$ with probability 1 for all $k = 1, 2, \ldots, n$ ($\varphi_0 \equiv 1$, as at time zero our random variables X_0 and φ_0 degenerate into constants).

Discussion.

(a) If we insist that $Q[X]$ must be a positive continuous linear functional then *all models* which are possible reduce to models on the vector φ.
(b) The following interpretation is natural
   - φ_k (taking the role of v^k in Val[·]) is called *stochastic discount function*.
   - φ is called *stochastic discount vector*.

2.2 Valuations at Time $t > 0$ ($Q[X \mid \mathcal{F}_t]$)

Observe: a) seen from time zero $Q[X \mid \mathcal{F}_t]$ is a random variable.
   b) seen at time t $Q[X \mid \mathcal{F}_t]$ is an observed value.
   Also for $t > 0$ we have a

Representation Theorem We must have

$$Q[X \mid \mathcal{F}_t] = \frac{1}{\varphi_t} \mathrm{E}\left[\sum_{k=0}^{n} \varphi X_k \mid \mathcal{F}_t\right] \tag{5}$$

$$= \sum_{k=0}^{t} \frac{\varphi_k}{\varphi_t} X_k + \mathrm{E}\left[\sum_{k=t+1}^{n} \frac{\varphi_k}{\varphi_t} X_k \mid \mathcal{F}_t\right] \tag{6}$$

$$= \mathrm{A}[X \mid \mathcal{F}_t] + \mathrm{R}[X \mid \mathcal{F}_t]$$

$$= \text{accumulated payments} + \text{prospective reserve}$$

Proof. The interested reader can find the proof in [1].

Discussion. Observe that we take expectations only over *discounting functions* but not over *capitalization functions* "hence avoiding the stochastic interest rate paradox" mentioned in the beginning of this chapter.

2.3 Modelling Principle

(A) As all stochastic models for interest rates can be described by the discount vector $\varphi = (\varphi_0, \varphi_1, \ldots, \varphi_n)$, modelling consists in the construction of a probability distribution of φ.

(B) From (A) one then derives $Q[X]$ and the stochastic process $(Q[X \mid \mathcal{F}_t])_{t \geq 0}$.

Remark. At this junction it might be helpful to remind the reader that for the standard continuous models for interest we have the analogous step (A):

Construct a probability measure either

– for the spot rate (one factor models)
– for the short and the long rate (two factor models)
– for the "whole yield curve"

Our reasoning in the discrete case seems in the spirit of "whole yield curves".

Replacement. It is often convenient to replace the φ_k-variables by "year to year" Y_j-variables. We define

$$\varphi_k = \prod_{j=1}^{k} Y_j \qquad \left(\varphi_0 = \prod_{j \in \emptyset} Y_j \equiv 1 \right)$$

Obviously all our formulae involving φ_k's can be rewritten in the "Y-language".
For example we have

$$Q[X \mid \mathcal{F}_t] + \sum_{k=0}^{t} \frac{1}{Y_{k+1} \cdot \ldots \cdot Y_t} X_k + E\left[\sum_{k=t+1}^{n} Y_{t+1} Y_{t+2} \ldots Y_k X_k \,\middle|\, \mathcal{F}_t \right] \qquad (7)$$

from which we can easily derive the *recursion formula*

$$E\left[Y_t \, Q[X \mid \mathcal{F}_t] \mid \mathcal{F}_{t-1} \right] = Q[X \mid \mathcal{F}_{t-1}] \qquad (M)$$

which can be also formulated as follows

$$\left(\varphi_t \, Q[X \mid \mathcal{F}_t] \right)_{t=0,1,2,\ldots,n} \quad \text{form a martingale} \qquad (M')$$

Discussion. (M) is very handy for numerical computations. (M') is theoretically useful: e.g. it allows to prove that there are no *arbitrage opportunities*.

3 Simple Payment Streams; Pure Financial Risk

Let me first apply our valuation technique to simple payment streams (streams with deterministic payments). These payment streams still have a stochastic valuation $Q[X \mid \mathcal{F}_t]$ different from the deterministic valuation $Val[X \mid \mathcal{F}_t]$ due to the "risk" in the φ-variables or equivalently the Y-variables). We call this the *financial risk*.

The building stone for all simple payment streams is the Zero Coupon Bond which we are going to discuss now.

3.1 Zero Coupon Bonds; Discounting Values

The Zero Coupon Bond (ZCB) paying 1 unit of money at time m is represented by the payment stream

$$I_m = (0, 0, \ldots, 0, 1, 0, \ldots, 0) \quad (m \leq n)$$

where 1 appears at the mth column.

We define $D_m = Q[I_M]$ as the Value of the ZCB at a time 0. For actuaries: This is identical with the Discount Factor $D_{0;m}$ (m back to 0).

We need also the Forward Values and the Forward Discounting Factors $D_{m-1;m}$, $D_{k;m}$ ($k < m$). We read: $D_{k;m}$: Forward Price of ZCB with maturity m made at time 0; to be paid at time k. Or: $D_{k;m}$ Forward Discount Factor of 1 unit at time m made at time 0; to be paid at time k
and must have

$$D_{k;m} = \frac{D_m}{D_k} \tag{8}$$

which can be seen from the fact that the payment of $D_k \cdot D_{k;m}$ and of D_m at time 0 both give you 1 at time m.

Finally we represent the value of the ZCB at time $k < m$

$$Q[I_m \mid \mathcal{F}_k] = V_k^{(m)} = E[Y_{k+1} \ldots \mid \mathcal{F}_k]$$

and we note that our basic recursion reduces to

$$E\left[Y_k V_k^{(m)} \mid \mathcal{F}_{k-1}\right] = V_{k-1}^{(m)} \tag{M}$$

or

$$\left(\varphi_k V_k^{(m)}\right)_{k=0,1,2,\ldots,n} \quad \text{form a martingale} \tag{M'}$$

3.2 Explicit Models for $Y = (Y_1, Y_2, \ldots, Y_n)$

There are as many models as you can imagine and there is no limit to your imagination. The following three models are just presented as possible examples. Whether they are reasonable and useful for practical purposes still needs to be tested.

Model I "Uncertainty of interest rates". This model expresses in an actuarial way, the fact that interest rates are fluctuating, similar to risks in a heterogeneous insurance portfolio.

Basic Formulae for Model I. Transform:

$$Y_k = \epsilon(1 - Z_k) + \delta Y_k = \epsilon + \Delta Z_k$$

where $0 < \epsilon < \delta \leq 1$ and $\Delta = \delta - \epsilon$. The Y_k are weighted averages of ϵ and δ, where Z_k represent the stochastic weights. $Z_k^{(M)}$ takes values $\frac{j}{M}$ $(j = 0, 1, 2, \ldots, M)$.

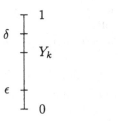

Remark: $M = 1$ leads to a binary model.

We define now the probability law of the vector Z with a pure Bayes construction as follows:

i) given p: $Z_k^{(M)}$ $(k = 1, 2, \ldots, n)$ are i.i.d with distribution

$$\frac{1}{M} \times \text{Binomial}(M, p).$$

ii) p has a Beta Distribution with parameters α, β

$$f_{\alpha,\beta}(p) = \frac{\Gamma(\alpha + \beta)}{\Gamma(\alpha) \cdot \Gamma(\beta)} p^{\alpha-1}(1 - p)^{\beta-1} \quad (0 \leq p \leq 1)$$

We derive:

(a)

$$D_n(\alpha, \beta) = \text{E}[Y_1 \cdot Y_2 \ldots Y_n] = \text{E}\big[(\epsilon + \Delta p)^n\big] = \sum_{k=0}^{n} \binom{n}{k} \epsilon^k \Delta^{n-k} \frac{\alpha^{[n-k]}}{(\alpha + \beta)^{[n-k]}}$$

$$(9)$$

with $\alpha^{[j]} = \alpha \cdot (\alpha + 1) \cdot \ldots \cdot (\alpha + j - 1)$ (factorial power)

(b) $V_k^{(n)} = E[Y_{k+1} \cdot Y_{k+2} \ldots Y_n \mid \mathcal{F}_k] = D_{n-k}(\alpha_k, \beta_k)$ with updating rules

$$\alpha_{k+1} = \alpha_k + MZ_k \quad (\alpha_0 = \alpha),$$
$$\beta_{k+1} = \beta_k + M(1 - Z_k) \quad (\beta_0 = \beta).$$

Observe:

1. The nice aspect of this model lies in the fact that all posterior distributions of p are again Beta distributions. Hence we can always work with the formula as derived under (a), in particular

$$V_k^{(n)} = D_{n-k}(\alpha_k, \beta_k) \quad \text{for all } k = 0, 1, 2, \ldots$$

2. the stochastic movement of $V_k^{(n)}$ is parameterized by the stochastic movement of the point (α_k, β_k) in the (α, β)-plane
3. The movement in the (α, β)-plane can be interpreted as *Polya's Urn*:
   - start with α white and β black balls.
   - after each trial add $Z_k M$ white and $(1 - Z_k)M$ black balls.
4. It is tempting to use this parameterization for more general models by changing the updating rules. This means that we keep the property that all posteriors of p are Beta distributions but relax our modelling assumptions in the spirit of time varying parameters.

Numerical results. For the numerical calculations the following parameter values have been used

$$M = 1 \quad \text{(binary model)}$$
$$\epsilon = 0.1$$
$$\delta = 1$$
$$\alpha = 4$$
$$\beta = 1$$

Table 1. Discount Factors D_1 to D_{20}

D_1 = 0.8200000000	D_{11} = 0.2893610988
D_2 = 0.6939999999	D_{12} = 0.2717020828
D_3 = 0.6011714286	D_{13} = 0.2560713358
D_4 = 0.5300585714	D_{14} = 0.2421388818
D_5 = 0.4738921429	D_{15} = 0.2296425433
D_6 = 0.4284335286	D_{16} = 0.2183714034
D_7 = 0.3909003155	D_{17} = 0.2081538756
D_8 = 0.3593933544	D_{18} = 0.1988489535
D_9 = 0.3315733860	D_{19} = 0.1903396964
D_{10} = 0.3094695276	D_{20} = 0.1825283081

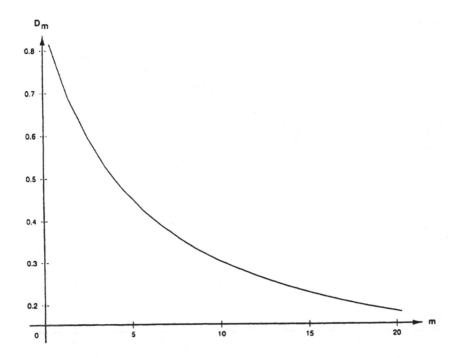

Fig. 1. Discount Factors D_1 to D_{20}

Table 2. Forward Yearly Discounts $D_{m-1;m}$, $m = 1, 2, \ldots, 20$

$D_{0,1}$ $= 0.8200000000$	$D_{10,11} = 0.9350229118$
$D_{1,2}$ $= 0.8463414633$	$D_{11,12} = 0.9389723910$
$D_{2,3}$ $= 0.8662412517$	$D_{12,13} = 0.9424710078$
$D_{3,4}$ $= 0.8817095194$	$D_{13,14} = 0.9455915128$
$D_{4,5}$ $= 0.8940373168$	$D_{14,15} = 0.9483918551$
$D_{5,6}$ $= 0.9040739228$	$D_{15,16} = 0.9509187638$
$D_{6,7}$ $= 0.9123943142$	$D_{16,17} = 0.9532103213$
$D_{7,8}$ $= 0.9193989878$	$D_{17,18} = 0.9552978676$
$D_{8,9}$ $= 0.9283743341$	$D_{18,19} = 0.9572074333$
$D_{9,10}$ $= 0.9305300443$	$D_{19,20} = 0.9589602030$

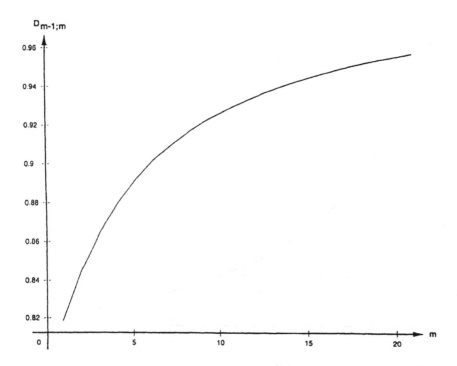

Fig. 2. Forward Yearly Discounts $D_{m-1;m}$, $m = 1, 2, \ldots, 20$

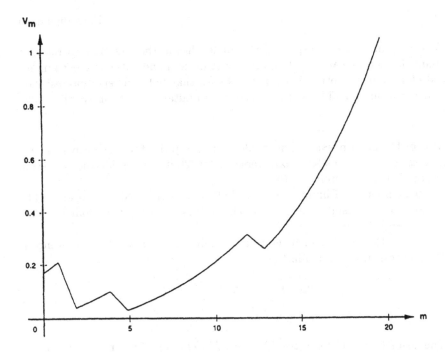

Fig. 3. Value of Zero Coupon Bond $V_k^{(20)}$, $k = 1, 2, \ldots, 20$. (single trajectory)

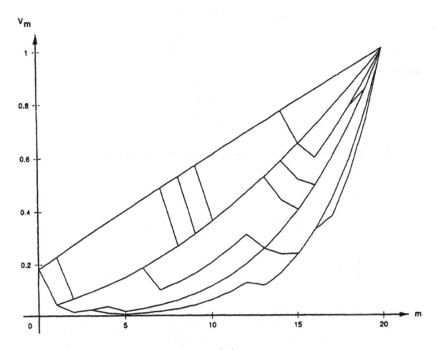

Fig. 4. Value of Zero Coupon Bond $V_k^{(20)}$, $k = 1, 2, \ldots, 20$. (20 superposed trajectories)

Discussion. The nice aspect of this model lies in the fact that all relevant quantities can be calculated in an explicit fashion and without great numerical effort. On the other hand Fig. 2 shows that only "inverse interest rate structures" are possible, which is of course a limitation of the use of the model I.

Model II "Ehrenfest". This model is in the spirit of some continuous time models as e.g. Vasicek [4], Cox-Ingersoll-Ross [2]. It uses the idea of Keynesian backwardation of interest rates.

Remember the Ehrenfest urn: $2s$ balls are distributed in two urns. The "random mechanism" is defined as follows: Pick one ball at random and shift it to the other urn.

If Y is the number of balls in urn number 1, we have the homogeneous Markov transition probabilities:

$$p_{y,y+1} = 1 - \frac{y}{2s} = \frac{1}{2} + \frac{s-y}{2s}$$

$$p_{y,y-1} = \frac{y}{2s} = \frac{1}{2} - \frac{s-y}{2s}.$$

This leads to the *Generalized Ehrenfest Model* (step of the movement $\pm\frac{1}{M}$):
For

$$p_{y,y+\frac{1}{M}} = \frac{1}{2} + a(b-y)$$

$$p_{y,y-\frac{1}{M}} = \frac{1}{2} - a(b-y)$$

we must have

$$y_{\max} = b + \frac{1}{2a}$$

$$y_{\min} = b - \frac{1}{2a}$$

Hence Y is defined over the range as indicated by the sketch:

Basic Formulae for Model II. The good method to calculate is—of course—by recursion (M)

$$V_k^{(n)}(y) = \mathrm{E}\left[Y_{k+1} V_{k+1}^{(n)}(Y_{k+1}) \mid Y_k = y\right] \qquad (M)$$

which becomes by homogeneity

$$D_{n-k}(y) = \mathrm{E}\left[Y_1 \cdot D_{n-k-1}(Y_1) \mid Y_0 = y\right]$$

or explicitly

$$D_{n-k}(y) = \left(y + \frac{1}{M}\right) D_{n-k-1}\left(y + \frac{1}{M}\right)\left[\frac{1}{2} + a(b - y)\right] \qquad (M'')$$
$$+ \left(y - \frac{1}{M}\right) D_{n-k-1}\left(y - \frac{1}{M}\right)\left[\frac{1}{2} - a(b - y)\right]$$

In the following table proceed from left to right using (M''):

Values for y	D_0	D_1	D_2
$y_{\max}$	1	:	:
$y_{\max} - \frac{1}{M}$	1	:	:
$y_{\max} - \frac{2}{M}$	1	:	:
:	:	:	:
$y + \frac{1}{M}$	:	:	:
y	1	$y + \frac{2a}{M}(b - y)$	:
$y - \frac{2}{M}$	:	:	:
:	:	:	:
$y_{\min} + \frac{1}{M}$	1	:	:
$y_{\min}$	1	:	:

Remark. This recursion method works for all homogeneous Markov Chain models for Y.

Numerical Results for Model II. For the numerical calculations the following parameter values have been used

$$\begin{array}{l} a = 5/2 \\ b = 16/20 \\ M = 20 \end{array} \quad \text{leading to} \quad \begin{array}{l} y_{\min} = 12/20 \\ y_{\max} = 1 \end{array}$$

Table 3. Discount Factors D_1 to D_{10}

y	12/20	13/20	14/20	15/20	16/20	17/20	18/20	19/20	20/20
D_1	0.6500	0.6875	0.7250	0.7625	0.8000	0.8375	0.8750	0.9125	0.9500
D_2	0.4469	0.4928	0.5406	0.5903	0.6419	0.6953	0.7506	0.8078	0.8669
D_3	0.3203	0.3646	0.4121	0.4629	0.5169	0.5743	0.6351	0.6995	0.7674
D_4	0.2370	0.2765	0.3196	0.3666	0.4176	0.4728	0.5322	0.5961	0.6645
D_5	0.1797	0.2135	0.2511	0.2927	0.3384	0.3884	0.4430	0.5022	0.5663
D_6	0.1389	0.1674	0.1994	0.2352	0.2750	0.3188	0.3670	0.4197	0.4772
D_7	0.1089	0.1327	0.1597	0.1900	0.2239	0.2616	0.3032	0.3493	0.3991
D_8	0.08647	0.1062	0.1287	0.1542	0.1827	0.2146	0.2500	0.2891	0.3320
D_9	0.06924	0.08558	0.1043	0.1255	0.1494	0.1762	0.2059	0.2389	0.2752
D_{10}	0.05585	0.06932	0.08480	0.1024	0.1223	0.1446	0.1695	0.1971	0.2276

Table 4. Forward Yearly Discounts $D_{m-1;m}$, $m = 1, 2, \ldots, 9$

y	12/20	13/20	14/20	15/20	16/20	17/20	18/20	19/20	20/20
$D_{0,1}$	0.6500	0.6875	0.7250	0.7625	0.8000	0.8375	0.8750	0.9125	0.9500
$D_{1,2}$	0.6875	0.7168	0.7457	0.7742	0.8024	0.8302	0.8578	0.8853	0.9125
$D_{2,3}$	0.7167	0.7399	0.7623	0.7842	0.8053	0.8260	0.8461	0.8659	0.8852
$D_{3,4}$	0.7399	0.7584	0.7755	0.7920	0.8079	0.8233	0.0380	0.8522	0.8659
$D_{4,5}$	0.7582	0.7722	0.7857	0.7984	0.8103	0.8215	0.8324	0.0425	0.8522
$D_{5,6}$	0.7730	0.7841	0.7941	0.8036	0.8126	0.8208	0.8284	0.8357	0.8427
$D_{6,7}$	0.7840	0.7927	0.8009	0.8078	0.8142	0.8206	0.8262	0.8315	0.8363
$D_{7,8}$	0.7940	0.8003	0.8059	0.8116	0.8160	0.8203	0.8245	0.8284	0.8310
$D_{8,9}$	0.8010	0.8558	0.8104	0.8139	0.8177	0.8211	0.8236	0.8246	0.8289

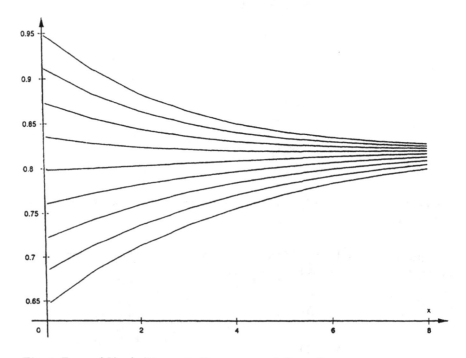

Fig. 5. Forward Yearly Discounts $D_{m-1;m}$, $m = 1, 2, \ldots, 9$

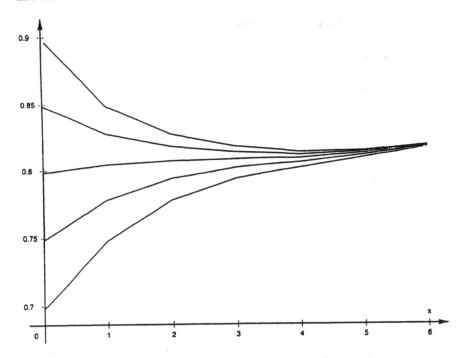

Fig. 6. Forward yearly discounts $D_{m-1;m}$ for a changed value of $M = 10$.

Discussion. Fig. 5 gives only "standard" and "inverse" term structures. If we change the parameter M, as done in Fig. 6, one receives also "mixed structures" where forward rates go first up and then down. As had to be conjectured this is completely analogous to the term structures obtained in the continuous case from the Cox-Ingersoll-Ross model.

Model III "Kalman". This model is in the spirit of evolutionary parameters as typically treated by the Kalman filter. Hence the name.

We assume $Y_k = e^{-\Delta_k}$ where $\Delta_k \sim \mathcal{N}(\mu_k, \sigma^2)$ where $\Delta_k = \mu_k + \epsilon_k$; $(\epsilon_k)_{k \in \mathbb{N}}$ i.i.d. $\mathcal{N}(0, \sigma^2)$.

The following three cases have been *proposed*:

(1) $\mu_j = \mu_{j-1} + \delta_j$ where δ_j i.i.d. $\sim \mathcal{N}(0, \tau^2)$, $\mu_0 \sim \mathcal{N}(\bar{\mu}_O, \Lambda_0)$
(2) $\mu_j = ab + (1-a)\mu_{j-1} + \delta_j$ where δ_j, μ_0 as in (a)
(3) as (b), with b stochastic i.e. $b \sim \mathcal{N}(\bar{b}_0, \eta^2)$

I have not yet made any computations based on model III.

4 Insurance Payment Streams

4.1 Basic Definitions

Remember the Valuation $Q[X]$ at time 0 (compare (4)):

$$Q[X] = \mathrm{E}\left[\sum_{k=0}^{n} \varphi_k X_k\right]$$

and the Valuation $Q[X \mid \mathcal{F}_t]$ at time $t > 0$ (compare (6) and (7)):

$$Q[X \mid \mathcal{F}_t] = \sum_{k=0}^{t} \frac{\varphi_k}{\varphi_t} X_k + \mathrm{E}\left[\sum_{k=t+1}^{n} \frac{\varphi_k}{\varphi_t} X_k \mid \mathcal{F}_t\right]$$
$$= A[X \mid \mathcal{F}_t] + \mathrm{R}[X \mid \mathcal{F}_t]$$
$$= \text{accumulated payments} + \text{prospective reserve}$$

$$Q[X \mid \mathcal{F}_t] = \sum_{k=0}^{t} \frac{1}{Y_{k+1} \cdot \ldots \cdot Y_t} X_k + \mathrm{E}\left[\sum_{k=t+1}^{n} Y_{t+1} Y_{t+2} \ldots Y_k X_k \mid \mathcal{F}_t\right]$$

Terminology. In these definitions we call:

- $X = (X_0, X_1, \ldots, X_n)$ *insurance variables or technical variables*
- $\varphi = (\varphi_1, \varphi_1, \ldots, \varphi_n)$ *financial variables*

(equivalently we also call the Y-variables *financial variables.*)

　　Assume that we have defined our model, i.e our probability distribution for

$$(X, \varphi) \quad \text{(which is a pair of two vectors)}$$

You should realize that all our basic definitions given in this section are *correct*, also if X and φ are dependent. On the other hand it seems natural for practical applications to try out first models where insurance variables and financial variables are independent, i.e. models in the style of Section 3 for φ and the traditional models called "mortality tables" for X.

　　You should also note that we always work under the assumption that the Equivalence Principle holds, i.e. $Q[X] = 0$ (The idea is that premiums of the insurer have been fixed such that $Q[X] = 0$).

　　Finally we want to stress that for the moment you should still keep in mind that we analyze the individual insurance policy. Only in section 4.4 we will address the aggregation problems of a portfolio of policies.

4.2 Annual Losses

As $Q[X] = 0$ we interpret

(a) $Q[X \mid \mathcal{F}_t]$ as *accumulated loss* (of insurer) *with interest* until time t

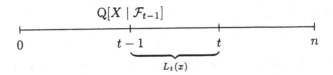

(b) $L_t(X) = Y_t\, Q[X \mid \mathcal{F}_t] - Q[X \mid \mathcal{F}_{t-1}]$ is then *the annual loss* from $t-1$ to t (discounted to the beginning of the time interval).

Observe:

$$M_m(X) = \sum_{k=1}^{m} \varphi_{k-1} L_k(x) = \varphi_m\, Q[X \mid \mathcal{F}_m] \tag{10}$$

The following lemma is obvious from (M').

Lemma 1. *The discounted sum of annual losses* $(M_m)_{m=1,2,\dots}$ *forms a martingale.*

Remark. For constant interest rates this lemma is a rewording of the theorem of Hattendorf [3] which has provoked a lot of controversies among older actuaries. Observe that we have proved in our context a much more general fact; namely that also in the case of stochastic interest rates discounted annual losses are increments of a martingale (hence e.g. *uncorrelated*). But we need more, namely a *decomposition* of annual losses into *technical loss* and *financial loss*. The main reason why this decomposition is needed is due to the fact that general insurance principles require that

– technical losses (and gains) must be *pooled* (law of large numbers), whereas
– the financial loss (or gain) essentially *belongs* to the individual policy.

4.3 Decomposition of Annual Losses

Remember: (Compare (10))

$$\underbrace{\sum_{k=m}^{m} \varphi_{k-1} L_k(x)}_{M_m} = \varphi_m\, Q[X \mid \mathcal{F}_m] = E\left[\sum_{j=0} \varphi_j X_j \mid \mathcal{F}_m\right]$$

The righthand side — as a sequence of conditional expectations of the same random variable with respect to the filtration $\mathcal{F}_m$ — is a martingale for any filtration! (for any increasing sequence of σ-algebras!)

It is natural to aim at a decomposition of the annual loss into technical and financial loss by construction a *finer filtration* sequence:

$$\mathcal{F}_0 \subset \mathcal{G}_1 \subset \mathcal{F}_1 \subset \mathcal{G}_2 \subset \mathcal{F}_2 \subset \mathcal{G}_3 \subset \mathcal{F}_3 \subset \ldots$$

with the following definitions

$$\mathcal{F}_m = \sigma(X_0, X_1, \ldots, X_m; \varphi_0, \varphi_1, \ldots, \varphi_m)$$

$\mathcal{F}_m$ as before, expressing the fact that at time m all X- and φ-variables are known up to and including number m.

$$\mathcal{G}_m = \sigma(X_0, X_1, \ldots, X_{m-1}; \varphi_0, \varphi_1, \ldots, \varphi_m)$$

$\mathcal{G}_m$ expressing the different fact where at time k the X-variables are only known up to and including $m-1$, whereas the φ-variables are known up to and including m.

We construct the new martingale sequence

$$M_0^{(\mathcal{F})}, M_1^{(\mathcal{G})}, M_1^{(\mathcal{F})}, M_2^{(\mathcal{G})}, M_2^{(\mathcal{F})}, M_3^{(\mathcal{G})}, \ldots$$

with respect to the filtration

$$\mathcal{F}_0 \subset \mathcal{G}_1 \subset \mathcal{F}_1 \subset \mathcal{G}_2 \subset \mathcal{F}_2 \subset \mathcal{G}_3 \subset \ldots$$

where we give the obvious interpretations

$$M_m^{(\mathcal{G})} = \mathrm{E}\left[\sum_{j=0}^{n} \varphi_j X_j \mid \mathcal{G}_m\right] = \varphi\, Q[X \mid \mathcal{G}_m] \tag{11}$$

$$= \sum_{k=0}^{m-1} \varphi_k X_k + \underbrace{\mathrm{E}\left[\sum_{k=m}^{n} \varphi_k X_k \mid \mathcal{G}_m\right]}_{\varphi_m R^+[X|\mathcal{g}_m]}$$

We use $R^+[X \mid \mathcal{G}_m]$ for the prospective reserve at time m *including the payment at time m* ("unknown" to $\mathcal{G}_m$)

Look at the martingale increases from time $m-1$ to time m:

$$M_m^{(\mathcal{F})} - M_m^{(\mathcal{G})} = \varphi_m\left[X_m + R[X \mid \mathcal{F}_m]\right] - \varphi_m R^+[X \mid \mathcal{G}_m]$$

$$M_m^{(\mathcal{G})} - M_{m-1}^{(\mathcal{F})} = \varphi_m R^+[X \mid \mathcal{G}_m] - \varphi_{m-1} R[X \mid \mathcal{F}_{m-1}]$$

$$\overline{M_m^{(\mathcal{F})} - M_{m-1}^{(\mathcal{F})} = \varphi_m\left[X_m + R[X \mid \mathcal{F}_m]\right] - \varphi_{m-1} R^+[X \mid \mathcal{F}_{m-1}]}$$

In analogy to the construction of the sequence $(M_m)_{m=1,2,\ldots}$ in Section 4.2 we obtain the corresponding annual losses by dividing above by φ_{m-1}; hence

Technical loss (financial basis unchanged)

$$(LT)_m = Y_m \big[X_m + R[X \mid \mathcal{F}_m] \big] - Y_m R^+ [X \mid \mathcal{G}_m] \tag{12}$$

Financial loss (claims experience identical)

$$(LF)_m = Y_m R^+ [X \mid \mathcal{G}_m] - R[X \mid \mathcal{F}_{m-1}] \tag{13}$$

Total loss

$$L_m = (LT)_m + (LF)_m \tag{14}$$

Which is our *decomposition*
   Observe: $L_m(X)$ is the same as in 4.2.

Remark. Observe that as $(LT)_m$ and $(LF)_m$ are related to a martingale; they have *each* expectation zero and are uncorrelated.

4.4 An Illustrative Example

The example is only used to illustrate the method. The model underlying our calculations for interest is however unrealistic (at least with the parameter values as chosen) Hence the reader is warned to use the example *only for the better understanding of the methodology.*
   Let us look at a Term Insurance for 3 years with constant premium P. Age at beginning is $x = 70$. Assume X and φ independent

$$\text{model for } X: \quad \begin{aligned} q_x &= 2.633\,\% \\ q_{x+1} &= 2.922\,\% \\ q_{x+2} &= 3.246\,\% \end{aligned}$$

Model for φ: as in 3.2 Model I (same parameters)
   We may represent the insurance risk by the following tree

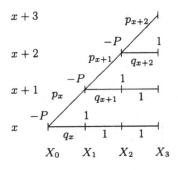

Notation. $X_k^{(n)} = \mathrm{E}[X_k \mid \mathcal{F}_n]$ ($= X_k$ for $n \geq k$).

Equivalence Principle $Q[X] = 0$. Hence,

$$0 = D_0(\alpha, \beta)X_0 \underbrace{+ D_1(\alpha, \beta)X_1^{(0)} + D_2(\alpha, \beta)X_2^{(0)} + D_3(\alpha, \beta)X_3^{(0)}}_{R[X|\mathcal{F}_0]}$$

One finds from this equation $P = 2.436\,\%$

Scheme for Sequence of Annual Losses

$$R[X \mid \mathcal{F}_0] = D_1(\alpha, \beta)X_1^{(0)} + D_2(\alpha, \beta)X_2^{(0)} + D_3(\alpha, \beta)X_3^{(0)}$$
$$Y_1 \overset{+}{R}[X \mid \mathcal{G}_1] = Y_1 X_1^{(0)} + Y_1 D_1(\alpha_1, \beta_1)X_2^{(0)} + Y_1 D_2(\alpha_1, \beta_1)X_3^{(0)}$$
$$Y_1[X_1 + R[X \mid \mathcal{F}_1]] = Y_1 X_1^{(1)} + \underbrace{Y_1 D_1(\alpha_1, \beta_1)X_2^{(1)} + Y_1 D_2(\alpha_1, \beta_1)X_3^{(1)}}_{Y_1\, R[X|\mathcal{F}_1]}$$

Sequence of Annual Losses

$(LF)_1$	$Y_1 = 1$	$+0.552\,\%$	← continue with $Y_1 = 1$
	$Y_1 = 0.1$	$-2.214\,\%$	
$(LT)_1$	alive	$-2.623\,\%$	← continue with "alive"
	dead	$+97.012\,\%$	
$(LF)_2$	$Y_2 = 1$	$+0.500\,\%$	← continue with $Y_2 = 1$
	$Y_2 = 0.1$	$-2.511\,\%$	
$(LT)_2$	alive	$-2.91\,\%$	← continue with "alive"
	dead	$+96.7\,\%$	
$(LF)_3$	$Y_3 = 1$	$+0.419\,\%$	← continue with $Y_2 = 1$
	$Y_3 = 0.1$	$-2.502\,\%$	
$(LT)_3$	alive	$-3.246\,\%$	← continue with "alive"
	dead	$+96.754\,\%$	

For the case $Y_1 = Y_2 = Y_3 = 1$ and "alive" at time 3 we have

$$\left. \begin{array}{l} (LF)_1 = 0.522\,\% \\ (LT)_1 = -2.623\,\% \end{array} \right\} \quad L_1 = -2.071\,\%$$

$$\left.\begin{array}{l}(LF)_2 = 0.500\,\% \\ (LT)_2 = -2.910\,\%\end{array}\right\} \quad L_2 = -2.410\,\%$$

$$\left.\begin{array}{l}(LF)_3 = 0.419\,\% \\ (LT)_3 = -3.246\,\%\end{array}\right\} \quad L_3 = -2.827\,\%$$

On average we have 2.436 % (which is the constant premium charged!). Observe that the Technical Gain is however bigger! It is needed to offset the financial loss!

4.5 Aggregate Annual Losses

All that we have discussed now is a model for the individual policy. Of course the decomposition of Annual Loss into Technical Loss and Financial Loss for the individual policy was made in order to allow aggregation. This is what I shall now discuss

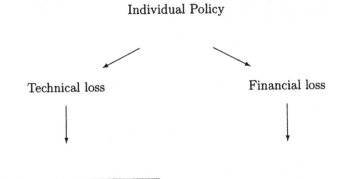

Individual Policy

Technical loss

Financial loss

Pot of Technical losses

Financial loss (or gain) belongs to policy

This "pot" has now to be analyzed similar to other insurance collectives.

Of course here we have a major problem if not the most important problem of Life Assurance. Let me call it The "AFIR Problem" of Life Assurance. (see below)

The AFIR Problem. As only gains can be transferred to the insured we have

instead of $-LF$ we guarantee to the insured
$(-LF)^+$

which is the *financial gain*
+ CALL OPTION

Two questions are immediate

1) How to price this CALL OPTION?
2) How to hedge this CALL OPTION?

I suggest these two topics as one of the most challenging research programs for the ACTUARY of the THIRD KIND.

References

1. Bühlmann, H: Stochastic discounting. Insurance Mathematics and Economics **11** No. 2 (1992) 113–127
2. Cox, J. C., Ingersoll, J. E. and Ross, St. A.: A Theory of the term structure of interest rates. Econometrica **53** No. 2 (1980) 385–407
3. Hattendorf, K.: Das Risiko bei der Lebensversicherung. Masius Rundschau der Versicherungen **18** (1868) 169–183
4. Vasicek, 0.: An equilibrium characterization of the term structure. Journal of Financial Economics **5** (1977) 177–188

Analyzing Default-Free Bond Markets by Diffusion Models

Franco Moriconi

University of Perugia

1 Introduction

In this paper we shall illustrate how diffusion models can be used to extract the term structure of interest rates from observed prices and to determine theoric prices and risk measures of term-structure derivatives. We shall assume throughout that the risk of default can be ignored.

We show first how a global analysis of the bond market can be performed using single factor diffusion models based on the no-arbitrage principle. As an example, we illustrate results of an application to Italian Government securities market, specializing the model as proposed by Cox, Ingersoll and Ross. The analysis will be confined to a period in which the assumptions of the model can be considered sufficiently realistic.

Secondly, we shall consider the effects of inflation uncertainty using a model with three state variables derived by Cox, Ingersoll and Ross within a general equilibrium setting. With this model it is possible to derive management strategies for portfolios containing both nominal and real bonds. Valuation of index linked securities and management of pension funds are straightforward applications. The choice of the model is motivated by the opinion that a general equilibrium model can be, for some application, more reliable and manageable than a pure arbitrage model.

2 One-factor Arbitrage Diffusion Models

Most of the one-factor diffusion models for interest rates based on the no-arbitrage principle are characterized by similar assumptions and properties.

Let us state the main assumptions on the bond market:

i) the market is perfect and frictionless. More precisely:
- all investors have homogeneous expectations;
- the market is competitive and each investor acts as a price taker;
- the market is open continuously and riskless istantaneous borrowing and lending takes place at the nominal interest rate $r(t)$;

- there are no taxes or transaction costs;

ii) riskless arbitrage opportunities are precluded;

iii) the state of the economy at time t is completely determined by the instantaneous interest rate $r(t)$, also called the spot rate, which represents the nominal yield of istantaneously maturing discount bonds.

2.1 The Basic Model

In this framework we shall assume that the spot rate is a Markov process with continuous paths, that is a diffusion process described by the Ito stochastic differential equation:

$$dr(t) = f(r(t), t) \, dt + g(r(t), t) \, dZ(t) \ ,$$

where $f(r(t), t)$ is the drift coefficient, $g^2(r(t), t)$ is the diffusion coefficient and $\{Z(t)\}$ is a standard Brownian motion.

Price dynamics. Let us consider an arbitrary interest rate sensitive (IRS) security, or an arbitrary portfolio of such securities. Generally, we shall represent it by a vector **c** which is assumed to contain all the contractual parameters. The price $V(t)$ at time t of the IRS financial claim **c** is a function of the spot rate at time t and of the time itself:

$$V(t) = V(r(t), t; \mathbf{c}) \ .$$

We shall assume throughout that the financial claim pays no dividends before a future date $t' > t$.

Using the well-known hedging argument, the no-arbitrage assumption implies that the price $V(t)$ must satisfy the second order partial differential equation:

$$V_t + (f + qg)V_r + \frac{1}{2}g^2 V_{rr} - rV = 0 \ , \tag{1}$$

where subscripts on V denote partial differentiation. The quantity $q(r(t), t)$ in this equation is a function independent of maturity called the *market price of risk* and is determined by the preferences of the investors. The presence of the taste dependent quantity q is one of the main features distinguishing the pricing of term structure derivatives from stock option pricing. The quantity:

$$\Phi(r(t), t) := -q(r(t), t) \, g(r(t), t)$$

is usually referred to as the *risk premium*, and $f^*(r(t), t) := f(r(t), t) - \Phi(r(t), t)$ is the so-called *risk-adjusted* drift.

Once the parameters of the spot rate process and the market price of risk are specified, the price of the IRS security is obtained by solving the fundamental valuation equation (1) subject to the appropriate boundary conditions.

Price of zero-coupon unit bonds. Let us denote by $v(r(t), t; s)$ the price at time t of a zero-coupon unit bond (ZCB) with maturity $s \geq t$. The model price of this elementary security is obtained by solving (1) under the terminal condition $v(r(s), s; s) = 1$.

The function $v(r(t), t; s)$ can be interpreted as the appropriate discount factor over the time interval $[t, s]$. As a function of s, it provides the term structure of ZCB prices prevailing on the market at time t. The term structure of interest rates can be readily obtained by evaluating the corresponding yield-to-maturity:

$$R(r(t), t; s) := -\frac{\log v(r(t), t; s)}{s - t} \ .$$

The price of deterministic payment streams:

$$\mathbf{x} := \{x_1, x_2, \ldots, x_m \ ; \ t_1, t_2, \ldots, t_m\}$$

can be immediately obtained by the linearity property of the price functional using the discount function $v(r(t), t; s)$:

$$V(r(t), t; \mathbf{x}) = \sum_{k=1}^{m} x_k \, v(r(t), t; t_k) \ .$$

The martingale property. In this no-arbitrage framework the most powerful result is the existence of an *equivalent martingale measure* for the price process. Referring to a bond maturing at time s, this property says that the discounted price process, defined as:

$$\left\{ V(T) \, e^{-\int_t^T r(u) \, du} \ ; \ t \leq T \leq s \right\} \ ,$$

is a martingale with respect to the risk-adjusted measure. A straightforward consequence is the integral representation for the price of contingent claims, also known as Feynman-Kac representation. Let us consider, for example, a stochastic ZCB which pays, at time s, the random amount $X_s = \eta(r(s))$, where η is a real-valued function fixed at time $t < s$. The time t price $V(r(t), t; X_s)$ of this bond is obtained as the solution of the fundamental valuation equation (1) under the terminal condition $V(r(s), s; X_s) = \eta(r(s))$ and can be expressed as:

$$V(r(t), t; X_s) = \widehat{\mathbf{E}}_t \left[\eta(r(s)) \, e^{-\int_t^s r(u) \, du} \right] , \tag{2}$$

where $\widehat{\mathbf{E}}_t$ is the expectation operator taken with respect to the equivalent martingale measure, conditional on the spot rate value observed at time t. Hence the bond price can be derived as the expectation of the discounted terminal payoff into an economy in which investor's expectations of the changes in the state variable are altered from f to $f + qg$.

Obviously, for a unit ZCB this expression simplifies to:

$$v(r(t), t; s) = \widehat{\mathbf{E}}_t \left[e^{-\int_t^s r(u) \, du} \right] \ .$$

Measures of interest rate risk. Applying Ito's lemma to the bond price $V(t)$ one immediately obtains that the percentage change in the price of a bond attributable to an unexpected shift in the spot rate is proportional to the semielasticity of the price with respect to r. Therefore the quantity:

$$\Omega(r(t), t; \mathbf{c}) = -\frac{V_r(r(t), t; \mathbf{c})}{V(r(t), t; \mathbf{c})}$$

provides a natural measure for the basis risk of the IRS security.

In many cases it is possible to express the interest rate risk in "natural" units of time by defining the *stochastic duration* of the claim $\mathbf{c}$ as the time-to-maturity of a ZCB with the same risk. Explicitly:

$$D(r(t), t; \mathbf{c}) = \varphi^{-1}\{\Omega(r(t), t; \mathbf{c})\} - t ,$$

where $\varphi^{-1}(\cdot)$ is the inverse function, with respect to s, of the ZCB risk measure:

$$\varphi(r(t), t; s) := -\frac{v_r(r(t), t; s)}{v(r(t), t; s)} .$$

The *stochastic immunization theory* is based on these risk measures [1].

The normalized prices. In order to represent all the bonds traded on the actual market into a unique picture we shall use the concept of normalized price. Let us briefly recall the definition (for more details see Castellani, De Felice and Moriconi, 1992).

By the stochastic immunization theorem any IRS assets $\mathbf{c}$ is equivalent to a unit ZCB having the same model price and the same duration. For any default-free bond $\mathbf{c}$ traded on the market we can observe the actual price $Q(t; \mathbf{c})$ and we can derive the model price $V(r(t), t; \mathbf{c})$ and the stochastic duration $D(r(t), t; \mathbf{c})$. Moreover we can also derive $v(r(t), t; t + D)$, that is the model price of the ZCB with the same duration.

The *normalized price* of $\mathbf{c}$ is then defined as:

$$\Pi(r(t), t; \mathbf{c}) := \frac{Q(t; \mathbf{c})}{V(r(t), t; \mathbf{c})} v(r(t), t; t + D) , \tag{3}$$

and can be interpreted as the market discount factor on the time horizon D.

Since $\Pi \geq v(D)$ if $Q \geq V$ (and vice versa), the rationale of the definition is to compare the market price of the bond with the equilibrium price of the equivalent unit ZCB. In the (D, Π) plane the function $\Pi = v(D)$ is the *market equilibrium line* prevailing at time t.

2.2 The One-factor Cox, Ingersoll and Ross Model

The most widely known model for the valuation of IRS securities is probably the single-factor diffusion model proposed by Cox, Ingersoll and Ross (CIR) (1985b). It can be derived by specializing the previous model as follows.

---

[1] For details and applications see De Felice and Moriconi (1991).

The spot rate dynamics. It is assumed that the istantaneous interest rate $r(t)$ follows a mean-reverting square-root process, described by:

$$dr(t) = \alpha[\gamma - r(t)]\, dt + \rho\sqrt{r(t)}\, dZ(t)\,, \qquad \alpha, \gamma, \rho > 0\,.$$

Under these dynamics it is expected that the interest rate will tend to revert to a long-term value γ with a speed of adjustment α proportional to its distance from γ. The diffusion term conforms to the empirical evidence that interest rate fluctuations are generally more pronounced when interest rates are high. Negative values of the spot rate are precluded and its transition density is noncentral chi-square.

The risk premium. The market price of risk is assumed to be:

$$q(r(t), t) = \pi\frac{\sqrt{r(t)}}{\rho}\,, \qquad \pi \text{ constant }.$$

Given the assumption on the form of the diffusion coefficient, the risk premium is given by:

$$\Phi(r(t), t) = -\pi\, r(t)\,.$$

So the risk adjustment is never unimportant or dominating with respect to the drift term.

Price and risk of unit ZCBs. Under these assumptions a closed-form solution of the valuation equation for the unit ZCB price can be derived. Denoting by $\tau := s - t$ the time-to-maturity of the bond one obtains:

$$v(r(t), t; t + \tau) = A(\tau)\, e^{-r(t)\, B(\tau)}\,, \tag{4}$$

where $A(\tau)$ and $B(\tau)$ are functions independent of r having the following form:

$$A(\tau) = \left\{\frac{2de^{(\alpha - \pi + d)\tau/2}}{(\alpha - \pi + d)[e^{d\tau} - 1] + 2d}\right\}^{2\alpha\gamma/\rho^2}\,, \tag{5}$$

$$B(\tau) = \frac{2\left[e^{d\tau} - 1\right]}{(\alpha - \pi + d)[e^{d\tau} - 1] + 2d}\,, \tag{6}$$

where:

$$d := \sqrt{(\alpha - \pi)^2 + 2\rho^2}\,.$$

For the applications, it is useful to observe that $A(\tau)$ and $B(\tau)$ are determined only by the risk-adjusted parameters $\alpha - \pi, \alpha\gamma$ and ρ. The function $B(\tau)$ also measures the interest rate risk of pure discount bonds having τ years to maturity.

Expected rates and term premia. The expression (4) for the time t price of unit ZCBs also allows the derivation of expected future spot prices:

$$\mathbf{E}_t[v(r(T), T; s)] = A(s - T)\, \mathbf{E}_t \left[e^{-r(T)\, B(s-T)} \right], \ t \leq T \leq s \ ,$$

using the expression for the conditional moment generating function of $r(T)$ provided by Feller (1951). If these expected prices are used to derive expectations of future interest rates, we can obtain an assessment of the *term premia* prevailing on the market at time t by simply computing the difference between expected spot rates and the forward rates implied in the current term structure.

Estimation procedures. In practical applications the parameters of the valuation model must be estimated from actual data. The most frequently used estimation procedures for the CIR model can be generally grouped into two main classes, two stage methods and cross-section methods.

With the *two stage methods* all the parameters of the model can be identified. In the first stage the three parameters of the spot rate process, α, γ and ρ, are estimated from a time series of observed short-term rates. The time series estimation is usually performed by a linear regression procedure based on a linear approximation of the discrete-time equivalent of the stochastic differential equation for $\{r(t)\}$. In the second stage the the estimated parameters from the first stage and the observed prices of bonds with different maturities are combined to estimate the preference dependent parameter π.

With the *cross-section method*, proposed by Brown and Dybvig (1986), the current value of $r(t)$ and the risk-adjusted parameters of the model, $\alpha - \pi$, $\alpha\gamma$ and ρ are simultaneously estimated using cross-sections of prices. Precisely, for a single day, the market prices of fixed rate bonds are observed and the estimated parameters are these producing the best least squares fit of model prices to observed market prices. Therefore the cross-section estimation of the function $v(r(t), t; s)$ given by (4) involves non-linear regression procedures.

The risk-adjusted parameters which may be obtained by cross-section methods are all we need in order to obtain arbitrage prices. However all the original parameters must be estimated if we need to derive expectations and probabilities.

3 Applications to the Italian Treasury Bond Market

As an example of application of the one-factor CIR model, we shall illustrate some results of a global analysis of the Italian Government securities market. The main classes of bonds currently traded on this market can be roughly described as follows:

- *Buoni Ordinari del Tesoro* (BOT) are ZCBs, issues twice a month, for maturities of three, six and twelve months. They represent about 30% of the marketable public debt.

- *Buoni del Tesoro Poliennali* (BTP) are fixed rate bond with semiannual coupons. In the recent years the authorities have stretched their maturity issuing 7 and 10 year bonds. The BTPs traded on the market amount to about 20% of the public debt.
- *Certificati di Credito del Tesoro* (CCT) are floating rate securities with semi-annual coupons indexed to BOT yield at issue. They are very popular securities, representing up to 40% of outstanding debt.
- *Certificati del Tesoro Optabili* (CTO) are fixed rate securities with semiannual coupons and typical maturity of six years at issue, with a put option to sell the bond at the face value at the end of the third year. They have been issued since December 1988 and represent about 5% of the marketable public debt.

A very liquid screen-based market exists for these securities (*Mercato Telematico*, MTS), in which some 30 market makers continuously quote bid-ask prices. Daily volumes easily reached 6,500 billions lire in 1992. All data in our analysis were taken from the MTS market.

3.1 A Time Series of Cross-sectional Estimates

In a first example of application, the risk-adjusted parameters and the spot rate value have been estimated by the cross-section method on the observed prices of fixed rate bonds (BTPs and BOTs) for each quotation day from January 9, 1990 to July 21, 1992. On each date the parameters estimated on the preceding quotation date were used as starting values in the non-linear regression procedure. The results are illustrated in Fig.1, in which only estimates on each tuesday are reported.

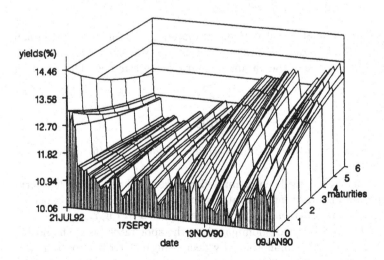

Fig. 1. Weekly evolution of the yield curves on the Italian Treasury bond market. Cross-sectional estimates

The yield curves were positively sloped until the end of 1990, progressively flattened during 1991 and reversed their slope starting from January 1992. Also in this case it appears to be confirmed the empirical regularity that the inversion of the yield curve predicts economic recession.

Two dramatic upward shifts in the term structure are evident from Fig.1 near the end of the observation period. This was the beginning of a period of instability. Even more dramatic shocks occurred in the following months. These effects resulted from a deep confidence crisis and changes in market expectations about the behaviour of policy makers. Therefore they can be related to a new element: the perception of default risk (the risk that the Government might not honour its debt). However, this second source of uncertainty is not included in our single-factor model. Here we are interested, instead, in the illustration of how we can represent all the different bonds traded at the same date into a unique picture. So we shall concentrate now on a single date picked out in the previous time period. Let us refer, for example, to the quotation date May 26, 1992. Also in this case the prices were observed on the MTS market. The parameter estimation was performed both by the two-stage method and by the cross-sectional method.

3.2 A Two-stage Estimate

In the first case the historical information was derived from the monthly time series of the auction yields of 3-month Treasury Bills, which were used as a proxy for the spot rate $r(t)$. This time series covers the period from January 1981 up to April 1992. The estimated values (on annual basis) were:

$$\alpha = 0.23552 \,, \ \gamma = 0.11156 \,, \ \rho^2 = 0.0030532 \,.$$

The risk premium parameter π was then estimated by a non-linear regression procedure on the BOT and BTP prices observed on May 26, 1992, with constraints given by the values of the parameters already determined. The resulting value for π was:

$$\pi = -0.022102 \,,$$

with a mean square error of the residuals equal to 0.24566 lire (for a face value 100 lire).

All the parameters having been estimated, we were able to derive the term structure of spot and forward interest rates, as well as the series of the expected future spot rates. The corresponding curves are reported in Fig.2.

The implied forward rates are dominated by the spot rates (as it should be, given that the yield curve is monotonically decreasing), but the expected spot rates are systematically higher. This implies negative term premia, with absolute value increasing with maturity.

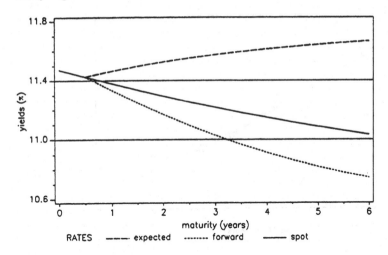

Fig. 2. Current rates and expected future spot rates on May 26, 1992. Two-stage estimate

3.3 A Global Analysis of the Treasury Bond Market on a Fixed Date

In order to provide a unified description of the bond market on the chosen date (May 26, 1992), we shall analyze the four main classes of Italian Treasury bonds using the definition (3) of normalized price. To represent all the bonds traded on the market on the (D, Π) plane only the risk-adjusted parameters of the CIR model are needed. Thus it is sufficient to perform a cross-sectional estimate on the observed prices of fixed rate bonds.

Analysis of fixed rate Treasury bonds. The cross-sectional estimation procedure applied on May 26, 1992 provided results consistent with those of the two-stage procedure. Twentythree BTPs and two BOTs were traded on the estimation date. Their position on the (D, Π) plane is illustrated in Fig.3.

The maturities (durations) of the two BOTs were 157 and 354 days. The BTP duration ranged between 1.79 and 4.32 years. The ability of the CIR model to fit the BOT and BTP actual prices is fairly good: the mean squared pricing error was about 1/4 lire, for a face value of 100 lire.

Analysis of variable rate Treasury bonds. As a second step of the analysis, the parameters estimated on the observed prices of fixed rate bonds were used to derive price and risk of CCTs. Almost one half of the Italian public debt is represented by Treasury bonds with stochastic cash flows, like variable rate bonds (CCTs) and fixed coupon bonds with embedded options (CTOs).

CCTs are medium or long-term Government securities in wich coupon payments adjust according to changes in short-term interest rates, since the coupon values are determined as a function of the issue prices of six or twelve-month Treasury Bills (BOTs). The indexation mechanism is such that only the next

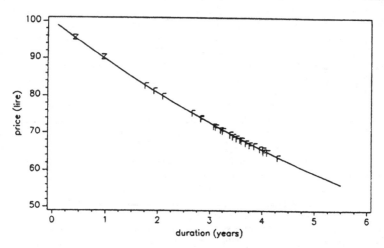

Fig. 3. Equilibrium line and normalized prices of fixed rate bonds. Z: Zero-coupon bond (BOT); F: Fixed-coupon bond (BTP)

two coupons, at most, are known at any valuation date (for more details see Castellani, De Felice and Moriconi, 1990).

The model price of each stochastic coupon must satisfy the fundamental valuation equation (1) under the appropriate boundary condition, which can be easily obtained using the integral representation (2) for the price of IRS securities. The resulting differential problem has a tractable closed-form solution (see Castellani, 1988). An accurate analysis of the indexation mechanism and of the self-immunization characteristics of CCTs suggests that holding one of these bonds should be substantially equivalent to rolling over a series of BOTs. Because of some imperfections in the indexation formula, such as a short time lag in the determination of the uncertain coupons, the stochastic duration should fluctuate around the maturity of the reference BOT during the life of the bonds. The results provided by the model are reported in Fig.4, where CCTs are represented in the (D, Π) plane.

Using as a reference the $D - \Pi$ relationship (the market equilibrium line) obtained from fixed rate securities,it is immediately apparent that CCTs are strongly underpriced by the market; the mean pricing error is about 3.4 lire (for 100 lire face value).

One possible explanation of this mispricing could be obtained by taking into account fiscal effects. In the estimation procedure we applied the model by considering the bond cash-flows net of the witholding tax. However different tax brackets can be identified on the Italian bond market. Adopting a simplifying view, we can refer to two different fiscal classes: tax-exempt and tax-payers investors. Owing to the indexation mechanism of CCTs, holding an indexed coupon could give a tax-exempt agent the possibility to invest at the pre-tax future market rate. If this possibility was perceived by the market as an effective arbitrage opportunity the stochastic coupons of CCTs should be eval-

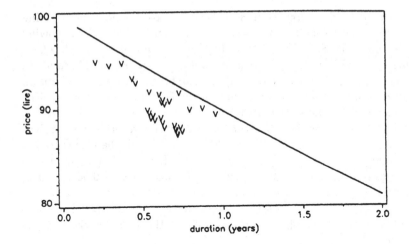

Fig. 4. Normalized prices of CCTs. V: Variable rate bond

uated as in a pre-tax framework. If the correction for this potential tax effect is introduced into the valuation model, the systematic mispricing of CCTs results to be eliminated, at least on some observation dates.

Another possible explanation of the CCT mispricing could be obtained by assuming that they are exposed to additional sources of risk. An important example may be the inability of the Treasury Bill issue rate, which is used as the reference rate in the determination of the indexed coupons, to correctly represent the short-term market rate. This would result into a kind of default risk if the market assumes as possible that the reference rate is forced below its equilibrium value by the issuer [2].

Multiple risk effects cannot be described by our single-factor model which assumes the interest rate risk as the unique source of uncertainty. However an empirical assessment of these effects can be obtained using the bond representation in the (D, Π) plane. The bond mispricing identified by the normalized price method is defined in connection with a time measure of risk. Hence the displacement of the security from the equilibrium line can also be interpreted in terms of a mispecification of the risk inherent to the bond. Measuring the displacement along the time axis we obtain, for any bond $\mathbf{c}$, a measure $\Delta(t; \mathbf{c})$ of additional risk implicitly defined by the equation:

$$\Pi(t; \mathbf{c}) = v(t; t + D - \Delta) \ .$$

As shown in Fig.4, the stochastic duration of CCTs derived by the single factor model ranges between 0.21 years and 1.16 years. If the pricing error of these bonds is interpreted as the risk premium asked by the market for facing additional sources of uncertainty, the implied additional risk is measured by an increment of the stochastic duration between 0.12 and 0.57 years.

---

[2] A systematic analysis of possible causes of CCT mispricing can be found in De Felice, Moriconi and Salvemini (1993).

Analysis of optable Treasury bonds. CTOs are fixed-coupon bonds giving to the holder the right to sell the bond at face value on a fixed date. Following Ananthanarayanan and Schwartz (1980) they can be modelled as *retractable bonds* (a long-term BTP and a european put option) or as *extendible bonds* (a short-term BTP and a european call option) and can be priced using bond option formulas currently available in the literature on the CIR model (see., e.g., CIR, 1985b, Jamshidian, 1987, Longstaff, 1990).

Using the parameters previously estimated from the fixed rate bond prices we derived the model price and the stochastic duration of all the CTOs traded on the MTS market on May 26, 1992. The optable Treasury bonds appear to be slightly underpriced by the market, with an average error of 0.96 lire. The duration range is between 1.74 and 3.15 years.

These results can be added to the previous pictures, as illustrated in Fig.5 which provides a nearly complete representation of the bond market on the estimation date.

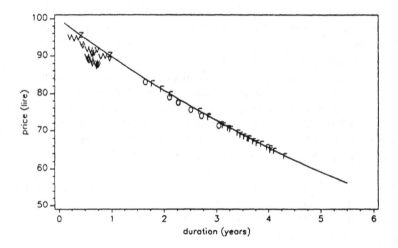

Fig. 5. Equilibrium line and normalized prices of Italian Treasury bonds. Z: Zero-coupon bond (BOT); F: Fixed-coupon bond (BTP); V: Variable rate bond (CCT); O: Optable bond (CTO)

4 A General Equilibrium Diffusion Model with Inflation Uncertainty

Usually, criticism on the one-factor CIR model concentrates on two main features: the long-term yield is constant over time and the istantaneous returns on bonds of all maturities are perfectly correlated: the latter property, at least, is clearly inconsistent with reality.

On the other hand, the one-factor CIR model is often considered more internally consistent by a theoric point of view with respect to pure arbitrage models, being derived from a general equilibrium model (CIR, 1985a) in which the interest rate process and the functional form of the market price of risk are endogenously derived and not simply "drawn out of the air". However it is not sufficiently pointed out (at least in practicionner's world) that the intertemporal general equilibrium model proposed by CIR describes a non monetary economy: payments are made in terms of one single consumption good and the borrowing/lending rate is a real interest rate. Using this model to describe nominal interest rates would be completely justified only in an oversimplyfied economy in which the price level, that is the money value of the consumption good, is constant over time (or at least has a deterministic evolution). In this case moving from real to nominal quantities would reduce to a trivial transformation of the value scale.

Leaving out these oversimplifications, it should be stressed that the use of the one-factor CIR model for pricing nominal bonds is not justified by general equilibrium considerations. So we are still using an arbitrage method for bond pricing which can be distinguished among competitors models only by properties of financial consistency (e.g. nonnegative interest rates) or of mathematical tractability (closed-form solution for some relevant pricing problems).

4.1 Toward General Equilibrium Models

Pricing models based on the no-arbitrage principle can be regarded as partial equilibrium models. In this framework the nominal interest rate process and the form of the risk-premium functions, one for each source of uncertainty, must be given exogenously. This is not completely satisfactory, however, because these quantities depend on the real economic activity, individual preferences and investor's attitude toward risk. As shown by CIR (1985b), the arbitrary choice of the functional form of the factor risk premiums can lead to internal inconsistencies or arbitrage opportunities.

The inadequacy of partial equilibrium models is even more evident when the securities to be valued have payoffs depending on real economic quantities, such as inflation, wages or price level. This issue is particularly relevant in the valuation and management of pension fund liabilities.

In the remaining part of this paper we shall show how both nominal and real bonds, like price index linked bonds with guaranteed minimum, can be correctly priced by a general equilibrium model with three state variables. This model was also proposed by CIR (1985b) in order to describe the effects of inflation uncertainty. When applied to the valuation of nominal bonds this framework reduces to a two-factor model derived independently by Richard (1978) using arbitrage methods [3].

---

[3] Also the two papers by CIR quoted as (1985a) and (1985b) were circulating from 1978 as a working paper. However the Richard model was derived independently. An application of Richard's results to the immunization of assets and liabilities of an

4.2 The Three-factor Model

Let us shortly describe the dynamics of the three state variables.

It is assumed that the *real istantaneous interest rate* $x(t)$ and the *expected istantaneous inflation rate* $y(t)$ follow mean-reverting square-root diffusion processes:

$$dx(t) = \alpha_x[\gamma_x - x(t)]\,dt + \rho_x\sqrt{x(t)}\,dZ_x(t)\,, \quad \alpha_x, \gamma_x, \rho_x > 0\,,$$

$$dy(t) = \alpha_y[\gamma_y - y(t)]\,dt + \rho_y\sqrt{y(t)}\,dZ_y(t)\,, \quad \alpha_y, \gamma_y, \rho_y > 0\,.$$

The third state variable $p(t)$ represents the *price level*; it is described by the stochastic differential equation:

$$dp(t) = p(t)\,y(t)\,dt + \rho_p\,p(t)\,\sqrt{y(t)}\,dZ_p(t)\,, \quad 0 < \rho_p < 1\,.$$

This dynamics implies that the expected percentage price change is equal to the anticipated inflation rate and the variability of the percentage price change is proportional to the anticipated inflation rate:

$$\mathbf{E}_t\left[\frac{dp(t)}{p(t)\,dt}\right] = y(t)\,, \quad \mathbf{Var}_t\left[\frac{dp(t)}{p(t)\,dt}\right] = \rho_p^2\,y(t)\,.$$

The two sources of nominal uncertainty, that is the Brownian motions $\{Z_y(t)\}$ and $\{Z_p(t)\}$, are assumed to be correlated:

$$\mathbf{Cov}_t\,[dZ_y(t), dZ_p(t)] = \rho^*\,dt\,, \quad \rho^*\text{ constant }.$$

However, real uncertainty is uncorrelated with nominal uncertainty:

$$\mathbf{Cov}_t\,[dZ_x(t), dZ_y(t)] = \mathbf{Cov}_t\,[dZ_x(t), dZ_p(t)] = 0\,.$$

Without going into details, we only recall that this model describes a real economy with a single good: all values are measured in terms of units of this good. The representative agent allocates his wealth seeking to maximize the expected utility of consumption over a fixed time horizon. Individual's utility is assumed to be logarithmic and the securities to be priced are not explicitly depending on wealth. The price $p(t)$ is simply a measure of value and has no effect on the underlying equilibrium. It simply represents the quantity of money needed for buying a unit of the consumption good at time t. Therefore in this framework money serves no other purpose than allowing to model nominal bonds.

In this general equilibrium setting the risk premium Φ relative to each state variable is determined as the covariance of changes in the state variable with percentage changes in the optimally allocated wealth W^*. Therefore the functional form of the risk premium $\Phi_x(t)$ relative to the real interest rate is endogenously

---

insurance company was illustrated by Boyle (1980) at the 21st International Congress of Actuaries.

determined. Namely $\Phi_x(t)$ results to be proportional to the current level of the real rate:

$$\Phi_x(t) = \mathbf{Cov}_t \left[\frac{dW^*(t)}{W^*(t)\,dt}, dx(t) \right] = -\pi\,x(t)\,, \quad \pi \text{ constant .} \tag{7}$$

Moreover the risk premia for the expected inflation and for the price level result to be completely specified. In the real economy framework, wealth is measured in real terms, so we obtain:

$$\Phi_y(t) = \Phi_p(t) = 0\,, \tag{8}$$

by the assumption of independence between real and nominal quantities.

4.3 Real Bonds

Let us denote by $B(t; \bar{X}_s)$ the real value at time t of a stochastic ZCB bond which pays at time $s > t$ the real amount $\bar{X}_s := \bar{\eta}(x(s), y(s), p(s))$, determined as a known function $\bar{\eta}$ of the values of x, y and p observed at time s. The model price of this security is obtained by solving the fundamental valuation equation:

$$\frac{1}{2}\rho_x^2 x\,B_{xx} + \frac{1}{2}\rho_y^2 y\,B_{yy} + \rho^*\rho_y\rho_p y\,p\,B_{yp} + \frac{1}{2}\rho_p^2 y\,p^2\,B_{pp} +$$
$$[\alpha_x(\gamma_x - x) + \pi x]\,B_x + \alpha_y(\gamma_y - y)\,B_y + y\,p\,B_p + B_t - \tag{9}$$
$$x\,B = 0$$

under the boundary condition $B(s; \bar{X}_s) = \bar{X}_s$. Let us introduce the second order differential operator $\widehat{\mathcal{D}}^R$, defined by:

$$\widehat{\mathcal{D}}^R B := \frac{1}{2}\rho_x^2 x\,B_{xx} + \frac{1}{2}\rho_y^2 y\,B_{yy} + \rho^*\rho_y\rho_p y\,p\,B_{yp} + \frac{1}{2}\rho_p^2 y\,p^2\,B_{pp} + \tag{10}$$
$$[\alpha_x(\gamma_x - x) + \pi x]\,B_x + \alpha_y(\gamma_y - y)\,B_y + y\,p\,B_p + B_t\,.$$

In this equation the coefficients of the first derivatives B_x, B_y and B_p are the drift terms adjusted for the risk premiums derived in the real framework and expressed by (7) and (8). Hence we can refer to $\widehat{\mathcal{D}}^R$ as the *real economy differential operator*. By this definition the valuation problem for $B(t; \bar{X}_s)$ can be written in the compact form:

$$\widehat{\mathcal{D}}^R B - x\,B = 0\,, \tag{11}$$
$$B(s; \bar{X}_s) = \bar{X}_s\,,$$

and its solution has the integral representation:

$$B(t; \bar{X}_s) = \widehat{\mathbf{E}}_t^R \left[\bar{X}_s\, e^{-\int_t^s x(u)\,du} \right]\,,$$

where $\widehat{\mathbf{E}}_t^R$ is the conditional expectation taken with respect to the risk-adjusted measure of the real economy.

Real unit ZCBs. If $\bar{X}_s$ is equal to a unit of consumption good we are considering a real unit ZCB with maturity s. Obviously the time t price $b(t;s)$ of this bond is obtained by solving (9) under the terminal condition $\bar{X}_s = 1$, but in this case the valuation equation can be simplified into a form similar to the valuation equation (1) of the one-factor CIR model; its solution is:

$$b(t;t+\tau) = A_1(\tau)\,e^{-x(t)\,B_1(\tau)}\,,\ \tau \geq 0\ ,\tag{12}$$

where the functions $A_1(\tau)$ and $B_1(\tau)$ are independent on x and have exactly the same form obtained in the single-factor model (equations (5) and (6)). They are determined by the risk-adjusted parameters of the process $\{x(t)\}$ only.

4.4 Nominal Bonds

In this framework nominal bonds can be modeled as contracts having payoffs whose real value depends on the price level. In particular, the real price $d(t;s)$ of a nominal unit ZCB with maturity s can be obtained by solving the valuation equation under the terminal condition $\bar{X}_s = 1/p(s)$, since having a quantity $1/p(s)$ of the consumption good at time s is equivalent to have a unit of money at the same date. It can be easily shown that the solution of this differential problem is:

$$d(t;s) = \frac{1}{p(t)}\,b(t;s)\,h(t;s)\ ,$$

where the function $h(t;s)$ has a form very similar to $b(t;s)$. Precisely:

$$h(t;t+\tau) = A_2(\tau)\,e^{-(1-\rho_p^2)y(t)\,B_2(\tau)}\,,\ \tau \geq 0\ ,$$

where the functions $A_2(\tau)$ and $B_2(\tau)$ are determined by the time-to-maturity τ and the by risk-adjusted parameters of the process $\{y(t)\}$ only and are completely analogous to the functions $A_1(\tau)$ and $B_1(\tau)$ in (12).

Nominal unit ZCBs. Now the nominal price $v(t;s)$ of a nominal unit ZCB is immediately obtained by multiplying the real price for the current price level $p(t)$; hence we have:

$$v(t;s) = b(t;s)\,h(t;s)\ .\tag{13}$$

Therefore, because of the independence between the real interest rate and the expected inflation rate, the nominal discount faxtor $v(t;s)$ is be expressed as the product of the real discount factor and the inflation discount factor. As a consequence, the nominal term structure, expressed in terms of yield-to maturity, can be represented as two superimposed term structures:

$$R(t;s) := -\frac{\log v(t;s)}{s-t} = -\frac{\log b(t;s)}{s-t} - \frac{\log h(t;s)}{s-t}\ .$$

This result, which expresses the nominal return as the real return on capital plus the opportunity cost of holding money, was also obtained by Richard (1978) in a partial equilibrium setting.

A straightforward implication of (13) is the following expression for the nominal spot rate $r(t)$ defined in Sec.2:

$$r(t) = x(t) + (1 - \rho_p^2)\, y(t) \ . \tag{14}$$

This relation, which expresses the nominal interest rate as a linear combination of the real interest rate and the expected rate of inflation, is the stochastic version of the celebrated *Fisher equation*. The factor $(1 - \rho_p^2)$, which is equal to one in the traditional version of the equation, is essentially a consequence of the Jensen inequality.

An example of cross-sectional estimate. As an example of application we estimated this model cross-sectionally on the prices of fixed rate nominal bonds observed on the same date. The two sets of risk-adjusted parameters $\rho_x, \alpha_x - \pi, \alpha_x \gamma_x$ and $\rho_y, \alpha_y, \alpha_y \gamma_y, \rho_p$ which specify the functions $b(t; s)$ and $h(t; s)$, respectively, were estimated by fitting through non-linear regression the function $v(t; s)$ given by (13) on the market prices of BTPs and BOTs quoted on January 29, 1991. The value of the variables $x(t)$ and $y(t)$ was constrained at a level x_0 and y_0, respectively. The value y_0 was determined by historical informations on the rate of inflation. The level of x_0 was derived residually using (14) after estimating the nominal spot rate $r(t)$ cross-sectionally by the one-factor CIR model.

The estimation results are illustrated in Fig.6 where the three yield curves relating to real rates, expected inflation rates and nominal rates are plotted. The nominal yield curve estimated by the one-factor model is also reported. The comparison between the two nominal curves confirms the intuition that the three-factor model provides a more wide variety of shapes.

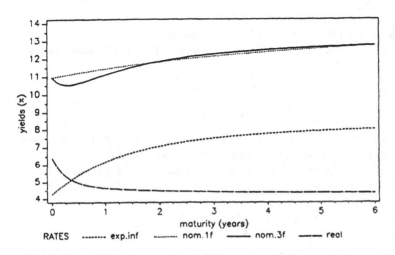

Fig. 6. Yield curves in the one and three-factor CIR model. Cross-sectional estimate on January 29, 1991

4.5 The Nominal Economy Setting

In the previous model it is also possible to derive a valuation equation in which all values are expressed in nominal terms. The crucial point is the determination of the form of the factor risk premia.

When we move to a nominal economy, wealth is expressed as $p(t) W(t)$. Hence the correlations of wealth with $y(t)$ and $p(t)$ are completely determined by the assumptions made on the volatilities of these two processes and on their correlation. The independence assumptions between real and nominal quantities leaves the premium Φ_x for the real interest rate risk unchanged while the other two risk premia result to be:

$$\Phi_y(t) := \mathbf{Cov}_t \left[\frac{d(p(t)W^*(t))}{p(t)W^*(t)\,dt}, dy(t) \right] = \rho^* \rho_y \rho_p\, y \tag{15}$$

and:

$$\Phi_p(t) := \mathbf{Cov}_t \left[\frac{d(p(t)W^*(t))}{p(t)W^*(t)\,dt}, dp(t) \right] = \rho_p^2\, y\, p \ . \tag{16}$$

The valuation equation for nominal bonds. Let us denote by $V(t; X_s)$ the nominal value at time t of a stochastic ZCB which pays at time $s > t$ the nominal amount $X_s := \eta(x(s), y(s), p(s))$. By Ito's lemma one can prove that in the nominal setting the valuation equation for this stochastic ZCB assumes the form:

$$\frac{1}{2}\rho_x^2 x\, V_{xx} + \frac{1}{2}\rho_y^2 y\, V_{yy} + \rho^* \rho_y \rho_p y\, p\, V_{yp} + \frac{1}{2}\rho_p^2 y\, p^2\, V_{pp} +$$
$$[\alpha_x(\gamma_x - x) + \pi x]\, V_x + [\alpha_y(\gamma_y - y) - \rho^* \rho_y \rho_p\, y]\, V_y + (1 - \rho_p^2)yp\, V_p + V_t -$$
$$r\,V = 0 \ , \tag{17}$$
$$V(s; X_s) = X_s \ .$$

where $r(t)$ is the nominal spot interest rate given by (14).

Also in this case we can write the valuation problem in the more compact form:

$$\widehat{\mathcal{D}}^N V - r\,V = 0 \,, \tag{18}$$
$$V(s; X_s) = X_s \,,$$

where $\widehat{\mathcal{D}}^N$ is the *nominal economy differential operator*, having the drift coefficients adjusted for the nominal risk premia given in (15) and (16).

Correspondingly, the Feynman-Kac representation for the price $V(t; X_s)$ of the nominal bond is given by:

$$V(t; X_s) = \widehat{\mathbf{E}}_t^N \left[X_s\, e^{-\int_t^s r(u)\,du} \right] \,,$$

where the conditional expectation $\widehat{\mathbf{E}}_t^N$ is now taken with respect to the risk-adjusted measure in the nominal economy.

If X_s is independent of p, then V is independent of the current price. In this case the valuation equation (19) simplifies and can be reduced to the equation derived by Richard (1978, p.48).

Letting $X_s = 1$ we obtain the alternative integral representation for the price of the nominal unit ZCB:

$$v(t; s) = \widehat{\mathbf{E}}_t^N \left[e^{-\int_t^s r(u)\, du} \right] \ .$$

Martingale properties. Both in the real and in the nominal economy there are fundamental martingale properties underpinning the integral representation of prices. Roughly stated these properties say that *the inflation-discounted price process is a martingale*. More precisely, we can prove that the process:

$$\left\{ p(T)\, e^{-\int_t^T y_u\, du} \ ; \ t \le T \le s \right\} \tag{19}$$

is a martingale with respect to the risk-adjusted measure in the real economy. Equivalently, the process:

$$\left\{ p(T)\, e^{-(1-\rho_p^2)\int_t^T y_u\, du} \ ; \ t \le T \le s \right\} \tag{20}$$

is a martingale with respect to the risk-adjusted measure in the nominal economy.

4.6 Index Linked Bonds with Minimun Guarantees

An interesting example of application of the previous model is the valuation of price index linked bonds with guaranteed minimum.

Let us consider a bond which pays at time s the nominal amount:

$$X_s := F \ \max\left(\frac{p(s)}{p(t)}, q \right) \ ,$$

where F is the face value, the ratio $p(s)/p(t)$ between the level price at maturity and the level price at the valuation date is the reference index determining the appreciation of F and q is the guaranteed minimum level for the appreciation factor.

Let $K = q\, p(t)$. We can write:

$$X_s = \frac{F}{p(t)} \left[K + \max\left(p(s) - K, 0 \right) \right] \ .$$

Hence, the nominal value of this bond can be expresed as:

$$V(t; X_s) = \frac{F}{p(t)} K\, v(t; s) + \frac{F}{p(t)} C(t; s, K) \ , \tag{21}$$

where $C(t; s, K)$ is the price of a european call option on $p(s)$ with exercise time s and strike price K. This price has the integral expression:

$$C(t; s, K) = b(t; s)\, \widehat{\mathbf{E}}_t^N \left[(p(s) - K)^+\, e^{-(1-\rho_p^2)\int_t^s y(u)\,du} \right] , \qquad (22)$$

which can be computed by numerical methods.

An alternative integral representation for the call price can be derived in the nominal setting; it results:

$$C(t; s, K) = p(t)\, b(t; s)\, \widehat{\mathbf{E}}_t^R \left[\frac{(p(s) - K)^+}{p(s)} \right] .$$

Comparison with alternative pricing models. The evaluation of price index linked bonds by the methods typically used for pricing commodity linked bonds can lead to systematic errors if the models are not corrected for some economic effects.

To illustrate this point let us pose, for simplicity, $K = 0$. In this case the previous bond effectively reduces to a real bond, since it guarantees to its holder the possibility of buying at time s the same amount of consumption good he or she can buy at time t with the face value F. Now (21) reduces to:

$$V(t; X_s) = \frac{F}{p(t)}\, V(t; p(s)) , \qquad (23)$$

where $V(t; p(s))$ is the nominal price of a bond which pays $p(s)$ at time s. Since the price of this bond, measured in terms of consumption good, is $b(t; s)$ we have:

$$V(t; p(s)) = p(t)\, b(t; s) . \qquad (24)$$

Hence:

$$V(t; X_s) = b(t; s)\, F . \qquad (25)$$

Therefore the value of the call option with vanishing strike price is given by the face value F, discounted by the real actualization factor.

Equation (24) can be derived more formally by the martingale property of the process (20). This property implies:

$$p(t) = \widehat{\mathbf{E}}_t^N \left[p(s)\, e^{-(1-\rho_p^2)\int_t^s y(u)\,du} \right] .$$

Thus (24) can be immediately obtained by observing that (22) leads to the expression:

$$V(t; p(s)) = b(t; s)\, \widehat{\mathbf{E}}_t^N \left[p(s)\, e^{-(1-\rho_p^2)\int_t^s y(u)\,du} \right] .$$

Let us consider the same pricing problem in the framework of tipical arbitrage models for *commodity linked bonds*, like that proposed by Schwartz (1982) (see also Carr, 1987). In these Black-Scholes type models the relevant martingale property implies:

$$p(t) = \widehat{\mathbf{E}}_t^* \left[p(s)\, e^{-\int_t^s r(u)\,du} \right] \; ,$$

where $\widehat{\mathbf{E}}_t^*$ is the expectation taken with respect to the risk-adjusted process of the nominal spot rate $r(t)$ and of the commodity price $p(t)$ [4]. Now the price $V^*(t; p(s))$ is expressed by:

$$V^*(t; p(s)) = \widehat{\mathbf{E}}_t^* \left[p(s)\, e^{-\int_t^s r(u)\,du} \right] \; , \tag{26}$$

and (23) gives:

$$V^*(t; X_s) = F \; .$$

If this result is used instead of (25) we are led to a systematic mispricing unless real interest rates uniformly vanish. We should have the same effect if the strike price was greater than zero.

This apparent inconsistency can be removed if we take into account the *service*, or *convenience*, *yield*. As pointed out by Ingersoll (1982, p.540), service yield "is a portion of the return on the commodity not reflected in the price change". If a service yield $\delta(t)$ is considered, the commodity return is given by $dp(t)/p(t) + \delta(t)\, dt$ and the relevant martingale property must be corrected as:

$$p(t) = \widehat{\mathbf{E}}_t^* \left[p(s)\, e^{-\int_t^s r(u)\,du} + \int_t^s \delta(u)\, e^{-\int_t^u r(z)\,dz}\, du \right] \; .$$

Hence (23) and (26) yield:

$$V^*(t; X_s) = F \, \frac{\widehat{\mathbf{E}}_t^* \left[p(s)\, e^{-\int_t^s r(u)\,du} \right]}{\widehat{\mathbf{E}}_t^* \left[p(s)\, e^{-\int_t^s r(u)\,du} + \int_t^s \delta(u)\, e^{-\int_t^u r(z)\,dz}\, du \right]} \; ,$$

which implies a reduction of the price comparable to that arising by discounting using the real interest rate [5].

This result indicates a logical equivalence between service yield and real return. Formally, we could incorporate the service yield into arbitrage models for commodity linked bonds by representing it as a dividend flow. However the service yield is not easy to be measured and modelled in a pure arbitrage framework.

---

[4] For example, in the traditional Black-Scholes model, with nominal interest rate $r(t)$ deterministic and constant, we have the well-known representation for the price of the underlying asset:

$$p(t) = e^{-r(t)(s-t)}\, \widehat{\mathbf{E}}_t^*[p(s)] \; .$$

[5] This argument is similar to the approximated method of adjusting for dividends in the Black-Scholes formula by reducing the stock price by the present value of the anticipated dividends.

In general equilibrium setting, instead, the service yield paid by the reference commodity is modelled in a natural way as the real return in the underlying economy.

References

Ananthanarayanan, A.L., Schwartz E.S.: Retractable and Extendible Bonds: The Canadian Experience. *The Journal of Finance* **35** (1980) 31–46

Boyle, P.P.: Recent Models of the Term Structure of Interest Rates with Actuarial Applications. *Transactions of the 21st Congress of Actuaries* (1980) 95–104

Brown, S.J., Dybvig, P.H.: The Empirical Implications of the Cox, Ingersoll, Ross Theory of the Term Structure of Interest Rates. *The Journal of Finance* **31** (1986) 617–630

Castellani, G.: Soluzione di equazioni differenziali del prezzo di titoli obbligazionari. *Note del Dipartimento di Scienze Attuariali e Matematica per le Decisioni Economiche e Finanziarie*, Università di Roma "La Sapienza", (1988)

Castellani, G., De Felice, M., Moriconi, F.: Price and Risk of Variable Rate Bonds: an Application of the Cox, Ingersoll, Ross Model to Italian Treasury Credit Certificates. *Proceedings of the 1st AFIR International Colloquium, Paris* **2** (1990) 287–310

Castellani, G., De Felice, M., Moriconi, F.: Single-factor Diffusion Models for the Global Analysis of the Italian Government Securities Market: Normalized Prices and Efficient Frontier. *Research Group on "Models of the Term Structure of Interest Rates", Working paper* 6 (1992)

Carr, P.: A Note on the Pricing of Commodity-Linked Bonds. *The Journal of Finance* **42** (1987) 1071–1076

Cox, J.C., Ingersoll, J.E., Ross, S.A.: An Intertemporal General Equilibrium Model of Asset Prices. *Econometrica* **53** (1985) 363–384

Cox, J.C., Ingersoll, J.E., Ross, S.A.: A Theory of the Term Structure of Interest Rates. *Econometrica* **53** (1985) 385–406

De Felice, M., Moriconi, F.: *La teoria dell'immunizzazione finanziaria. Modelli e strategie.* Il Mulino, Bologna (1991)

De Felice, M., Moriconi, F., Salvemini, M.T.: Italian Treasury Credit Certficates (CCTs): Theory, Practice and Quirks. *Banca Nazionale del Lavoro Quarterly Review* 185 (1993) 127–168

Feller, W.: Two Singular Diffusion Problems. *Annals of Mathematics* **54** (1951) 173–182

Ingersoll, J.E.: Discussion. *The Journal of Finance* **37** (1982) 540–541

Jamshidian, F.: Pricing of Contingent Claims in the One Factor Term Structure Model. *Merryl Lynch Capital Markets, Working paper* (1987)

Longstaff F.A.: The Valuation of Options on Coupon Bonds. *The Ohio State University, Working paper* (1990)

Richard, S.F.: An Arbitrage Model of the Term Structure of Interest Rates. *Journal of Financial Economics* **6** (1978) 33–57

Schwartz, E.: The Pricing of Commodity-Linked Bonds. *The Journal of Finance* **37** (1982) 525–539

Risk-based Capital for Financial Institutions

Phelim P. Boyle[1,2]

[1] Department of Finance, University of Illinois at Urbana Champaign
340 Commerce West, 1206 South Sixth Street, Champaign, Illinois 61820, U.S.A.
[2] School of Accountancy, University of Waterloo, Waterloo, Ontario, N2L 3G1 Canada

1 Introduction, Background and Overview

The issue of risk-based capital for financial institutions is of considerable current interest. We discuss how the option pricing paradigm provides a theoretical framework for the analysis of this topic. The modern approach to option pricing which was inspired by the seminal work of Black and Scholes has revolutionized our approach to both the theory and practice of many aspects of investment finance and corporate finance.

We start with a brief overview of the option pricing model. The aim is to highlight the key economic intuitions on which it rests. Derivations of the option model often use elegant but highly technical mathematics. These approaches provide a rigorous basis for the theory and are of fundamental importance. However such approaches can obscure the simple economic concepts which underpin the model. Accordingly we explore the derivation of the basic model in a discrete time model. This permits us to gain an uncluttered view of the most important economic intuitions in the pricing of derivative securities. These are: the principle of no-arbitrage and the concept of complete markets. The genesis of this discrete time approach stems from Arrow's beautiful 1952 paper.

We discuss a practical application that is of current interest. This concerns the determination of risk-based capital to cover credit risk and interest rate risk for a financial institution. In many countries the supervisory authorities and firms in the financial services industry have become concerned with this issue. In both Canada and the United States new risk based capital regulations have been drawn up for insurance companies and similar regulations prevail among the countries of the European Economic Community and elsewhere. In the United States the NAIC has proposed risk based capital requirements to cover the following types of risk: asset default risk; insurance risk; interest rate risk and business risk. These regulations specify the minimum amounts of capital to cover the different types of risk.

In the banking sector there is a prevailing emphasis on the coordination of regulation at the international level. The Bank for International Settlements

(BIS) [3] has taken an active role in formulating international banking standards and in 1988 proposed new risk-based capital requirements to cover credit risk. This agreement (sometimes called the Basle Accord) was fully implemented in 1922. It stipulates the minimum capital requirements to cover the credit risk of the bank's assets and off-balance sheet positions. The Accord relates principally to the institution's credit risk. Currently a working group of the BIS composed of representatives of the supervisory authorities from the twelve member countries is studying how a risk based capital formula could be devised to cover interest rate risk. The idea is that institutions with higher levels of interest rate risk will be compelled to allocate additional capital. At present two different approaches to interest rate risk in the case of deposit institutions have been developed by regulators in the United States. These agencies are the Federal Reserve Board (FRB) which supervises banks and the Office of Thrift Supervision (OTS) which regulates savings and loan institutions. These approaches [4] give specific details on the measurement of interest rate risk and the computation of the appropriate amount of risk based capital.

In the search for optimal capital standards there is a natural trade-off between theoretical accuracy and ease of administration. Given the complexity of the problem it is not surprising that the approaches adopted are often fragmented and sacrifice theoretical precision for pragmatic solutions. This is evident in the regulations that have been developed for both insurance companies and banks. From a policy perspective it would seem useful to have a theoretical framework within which to conduct a more integrated analysis. In this paper we summarize some research which takes some preliminary steps in this direction. While we focus on the banking industry our approach should be broadly applicable to any generic financial institution. We explore the possibility of using the option pricing paradigm as a vehicle for studying broader policy issues concerning the coordinated treatment of different types of risk in the determination of risk based capital standards. We build on the basic Merton model which views the common shares of a bank as a call option to purchase the assets of the bank from the depositors and discuss extensions of this model. In particular interest rate risk and credit risk can be analyzed simultaneously in one model. A major conclusion is that credit risk and interest rate risk are not simply additive. Hence it is incorrect to compute the additional capital for interest rate risk and add it to the existing credit risk amount as recommended by the existing proposals. However for pragmatic reasons this may be the approach adopted by the BIS. Many countries operate a deposit insurance plan or guaranty scheme to protect consumers in the case of insolvency. The option pricing approach also demonstrates the important connection between the optimal capital levels and the level of the insurance premiums for deposit insurance.

---

[3] The BIS Committee consists of representatives of twelve leading industrialized nations. These include Japan, The United States, Canada and nine western European countries.

[4] The FRB approach is described by Houpt and Embersit (1991) while the OTS approach is described by Transmittal Number 63, Office of Thrift Supervision (1992).

2 The Option Pricing Model

In this section we discuss a simple derivation of the option pricing model. We start with a one-period discrete time model. The relationship between the derivation of the model and the underlying economic concepts becomes more transparent in this setting. The one-period model can be extended to a multi-period model and ultimately to the continuous time Black Scholes model. The option pricing model prices a contingent pay-off in terms of currently traded assets. Indeed the model furnishes the current weights of the portfolio of the constituent assets that would ultimately replicate the target pay-off at maturity. It is important to note that the option model is a relative valuation model. It is silent on the issue of how the prices of the basic traded assets are obtained but it yields the current price of the option in terms of these assets.

Assume that at the end of the period there are two possible states of the world. The probability of each state is positive. Assume a frictionless market [5] with two traded securities. One is a risky security whose payoffs differ in the two states: the other is a riskless bond with identical payoffs in each state. We normalize the current prices of one share of each security to be unity at the start of the period. A single share of the risky asset is worth **u** in one of the states and **d** in the other. This convention differentiates the states as the up-state and the down-state. A unit of the riskless bond is worth **R** in each state. So far no assumption has been made about the relationship among **u**, **d** and **R**. The payoff matrix in this case is the two by two matrix with (**u, d**) in the first column and (**R, R**) in the second. It is now convenient to introduce the concept of **market completeness**. The intuitive concept is that the market is complete if any desired payoff is attainable. This means that the desired payoff pattern can be spanned by the existing securities. In the present example the market will be complete if the rank of the payoff matrix is equal to the number of state. This will be the case if **u** is not equal to **d** so that the determinant is non-zero. We assume without loss of generality that **u** is greater than **d**.

Our next concept the **no arbitrage principle** lies at the very foundation of modern asset pricing theories. A market does not permit arbitrage profits if it is impossible to make a riskless profit with a zero net investment. We expect well functioning markets to exhibit this property. One consequence of this principle is that if two securities have the same pay-offs in all future states they must perforce trade at the same current prices. The no-arbitrage principle imposes additional restrictions on **u**, **d** and **R**. We can demonstrate that **R** cannot exceed (or equal) **u**. If it did we could generate arbitrage profits. In the same way **R** cannot be

---

[5] It is important to stress that the financial economist's approach to the modelling of derivative securities places enormous importance on the existence of the market. This often contrasts with the traditional actuarial approach. As an example take the approach to interest rates. The financial economist's starting point will be the market prices for default free bonds. The more traditional actuarial approaches often start with the interest rate.

less than (or equal to) **d**. Hence we must have

$$u > R > d \tag{1}$$

Given that the market is complete we can find a combination of the risky asset and the bond that pays off one unit in the up-state and zero in the down-state and another combination that pays off one unit in the down-state and zero in the up-state. These primitive securities are known as Arrow-Debreu securities and let us denote their current prices by ϕ_u and ϕ_d. It is easy to obtain their explicit expressions.

$$\phi_u = \frac{R-d}{R(u-d)} \qquad \phi_d = \frac{u-R}{R(u-d)} \, . \tag{2}$$

The no arbitrage condition, (equation (1)), implies that these prices are positive. A portfolio which consists of the two Arrow Debreu securities pays off one unit in each state and thus corresponds to the one-period pure discount bond. Therefore its current price should be $1/R$. We see that the prices of the Arrow Debreu securities in equation (2) satisfy this relation. Furthermore if we multiply the Arrow Debreu prices by **R** we obtain two positive quantities which sum to unity. These have a natural interpretation as probabilities. Define

$$q = \frac{R-d}{u-d} \, ; \qquad 1-q = \frac{u-R}{u-d} \, . \tag{3}$$

It is natural to associate **q** with the up-state and $(1 - \mathbf{q})$ with the down-state. We label this probability measure as the **Q measure**. It is sometimes called the **equivalent martingale measure** because under this measure the price process for the risky asset suitably normalized is a martingale [6]. The equivalence between the existence of the **equivalent martingale measure** and the absence of arbitrage represents one of the most fundamental results in modern financial economics.

The valuation of any payoff in this model can be accomplished in terms of the Arrow Debreu securities. Suppose we have a European call option with strike price **K**. Denote the end-of-period call values as follows:

Event	Call Value
Up-state occurs	C_u
Down-state occurs	C_d

---

[6] Assume that the risky asset's price is expressed in units of the riskless asset. The current value of this ratio is unity. At the end of the period the expected value of this ratio under the Q measure is:

$$q\frac{u}{R} + (1-q)\frac{d}{R} = 1 \, .$$

Hence the normalized price process is a martingale under the Q measure.

In order to prevent arbitrage the current call price must be equal to

$$\phi_u C_u + \phi_d C_d \tag{4}$$

The current call price also can be written in terms of the equivalent martingale measure:

$$\frac{qC_u + (1-q)C_d}{R} = \frac{E_Q(C)}{R}. \tag{5}$$

This is the usual form of the discrete time option pricing formula as developed by Cox Ross and Rubinstein (1979). They illustrate how to extend it to the multi-period case. In the multi-period model the ability to trade at each period still permits us to attain any desired payoff pattern so that the market is still essentially complete. The same two basic economic intuitions of no arbitrage and market completeness provide the basis for the pricing formula in the multi-period setting.

The continuous time option pricing model, first derived by Black and Scholes can now be reached by two different routes [7]. We can take the continuous time limit of the multi-period model as in Cox Ross and Rubinstein (1979). Alternatively the model can be derived directly from first principles by using the machinery of stochastic calculus. It is important to stress that the latter approach relies also on the no arbitrage principle and market completeness. While the mathematics is much more technical the derivation rests on the same compelling intuitions that where used to derive the discrete time model. The Black Scholes formula assumes that the returns on the risky asset follow a geometric Wiener process with constant variance.

We use the following notation

S : Current price of underlying risky asset
T : Time until option maturity
r : Riskless interest rate (continuous compounding)
K : Strike price of option
σ^2 Variance of return on underlying asset
EC Black Scholes price of European call option.

The Black Scholes formula is:

$$EC = SN(d_1) - Ke^{-rT}N(d_2), \tag{6}$$

where $N(\ )$ is the cumulative normal density function and

$$d_1 = \frac{\log_e\{S/K\} + \left(r + \frac{\sigma^2}{2}\right)T}{\sigma\sqrt{T}}, \qquad d_2 = d_1 - \sigma\sqrt{T}.$$

The Black Scholes price of the European put may be obtained from put-call parity.

---

[7] Darrell Duffie (1988) gives a number of different rigorous derivations of the Black Scholes formula in his book.

One of the reasons that the option paradigm is so powerful is that it can be used to model not only traded option contracts but also a wide variety of contingent payoffs. Indeed many corporate liabilities can be modelled in this way since they represent claims on the underlying assets of the firm. Consider a firm whose capital structure consists of common shares and a single tranche of zero coupon debt with face value F. Upon bond maturity the shareholders are entitled to the value of the firm less the face value of the debt if this quantity is positive. The bondholders receive F if the firm value at maturity exceeds F; otherwise they own the assets of the firm. The common stock of the firm can be viewed as a call option to purchase the assets of the firm from the bondholders at a strike price of F. The bondholders' claim can be viewed as a long position in a riskless bond with face value F together with a short put option on the firm's assets with strike price F. While the actual boundary conditions are more complicated than this, the model is capable of providing useful insights.

The option paradigm provides a clear illustration of an important conflict-of interest situation between the bondholder and the stocholders of a firm. The stockholders claim becomes more valuable as the variance of the assets increases. This is because a call option is an increasing function of the volatility of the underlying asset [8]. Hence there is an incentive for the stockholders to increase the volatility of the firm's assets. Note that this can be accomplished without increasing firm value. If the shareholders' wealth is increased in this manner it must be at the expense of the bondholders. We can also see this by noting that the bondholders claim contains a short put position and since the put option's value also increases with increases in the volatility the value of the bondholders' claim falls with increases in the volatility. Of course there are usually are institutional and contractual arrangements in place to curb this tendency. These can take the form of bond covenants and specific legislation.

3 Risk Based Capital for Banks

The option model provides a convenient tool for analyzing the relationship between different types of risk and capital adequacy standards for deposit taking institutions. The Black Scholes option valuation model provides the equilibrium value for the common stock of a bank and also the equilibrium premium for deposit insurance. In this section we examine current issues in the area of risk-based capital for banks. We also provide background on the current BIS regulations for risk-based capital and describe models that have been proposed to deal with interest rate risk. The option framework permits an integrated approach to these topics.

The original stimulus for this line of research is due to Robert Merton (1977) who applied the Black Scholes option valuation model to value the common stock

---

[8] It can be verified by taking the partial derivative of the Black Scholes formula with respect to σ that both the call price and the put price are increasing functions of the volatility.

of a bank. Based on simplifying [9] assumptions the liabilities of the bank can be regarded as maturing at a fixed date. When the liabilities mature the equity holders are entitled to the excess (if any) of the asset value over the value of the liabilities. If the assets are insufficient to meet the liabilities the equity holders receive nothing. This payoff pattern is exactly the same as that of a call option where the strike price corresponds to the liability payment. Hence the current value of the equity can be obtained from the call option formula. In the case of insured deposits the insurer covers any shortfall between the value of the assets and the value of the deposits in the event of bankruptcy.

It is convenient to use the following notation:

T Time to next audit of bank's assets: "Maturity" of option

$V(0)$ Current market value of bank's assets

F Face value of bank's deposits

σ Standard deviation of rate of return on the value of the bank's assets

$E(0)$ Current value of bank's equity according to the model

$I(0)$ Current value of deposit insurance contract according to the model

From equation (6) the current value of the bank's equity is equal to:

$$E(0) = V(0)N(d_1) - Fe^{rT}N(d_2) \tag{7}$$

where $\mathbf{N}(.)$ and the $\mathbf{d}$'s are defined as before.

The current value of the deposit insurance premium is equal to:

$$I(0) = F\exp(-rT)N(-d_2) - V(0)N(-d_1), \tag{8}$$

The volatility, σ can be taken as a proxy for credit risk. Increases in σ tend to increase both the value of the bank's equity and the value of the deposit insurance. We can verify this from the comparative statics. The partial derivatives of E and I with respect to σ are both positive. This is the origin of the moral hazard problem. Shareholder wealth increases if $\mathbf{E}$ increases and if the deposit insurance is on a flat fee system the additional cost of the insurance is absorbed by the government insurer. There is an incentive for the owners of the bank to increase the asset risk. This phenomenon is dramatically illustrated by the recent history of the savings and loan industry in the United States.

The basic Merton model can also provide insights on the bank's decision with regard to capital. Additions to capital increase the market value of bank equity by the partial derivative of $\mathbf{E}$ with respect to $\mathbf{V}$ and hence the net increase in the owners' wealth is:

$$\frac{\partial E}{\partial V} - 1 = N(d_1) - 1. \tag{9}$$

---

[9] The assumption is that the maturity of the debt is the time until the next audit.

This expression is negative since $N(d_1)$ is less than one. Hence the bank's shareholders have no incentive to add capital [10]: quite the reverse.

The moral hazard story presented above does not incorporate the consequences of bankruptcy on the bank's shareholders. There are two ways of taking this into account. Marcus (1984) noted that the loss of charter value in the case of bankruptcy imposes costs on shareholders and that when this is taken into account in the Merton option pricing model the incentives can be reversed for a range of parameter values. Specifically it may be optimal for banks to increase their capitalization and it may also be optimal for well capitalized banks to pursue low risk strategies. On the other hand banks that have experienced severe loan losses and that are in poorly capitalized positions may have a strong incentive to pursue very risky strategies.

An alternate approach that obtains similar results is to assume that the supervisory agency imposes regulation to restrain the bank's risk taking behavior (see Buser, Chen and Kane (1981)). This regulation often consists of an imposed capital requirement. This regulatory cost can be structured in terms of a put option held by the insurer (see Kendall (1991)). The equity holders lose the minimum capital in certain states of the world and this cost influences their behavior in much the same way as the existence of a charter value. This approach also produces situations where it is rational for banks to increase their capital and also where it is rational for banks to control the level of the riskiness of their asset portfolios.

Such a model provides a framework to analyze asset risk, minimum capital standards and a stylized version of optimal bank behavior. It can be used to analyze capital adequacy regulations such as the existing BIS guidelines for credit risk. We can derive the equilibrium relationship between the asset risk: $-\sigma$ and the optimal capital standards. Within this model it is assumed that deposit insurance is fairly priced: in other words that the premium charged accurately reflects the risk. The model also suggests that the asset risk should be computed by taking the aggregate risk of the entire asset portfolio into account. The current BIS approach to credit risk is somewhat piecemeal since it is based on individual risk weights for different asset classes and thus the BIS regulations deviate from this ideal.

Currently a BIS committee is working on proposals for risk-based capital to cover interest rate risk. Interest rate risk is the risk that changes in the level of market interest rates might adversely impact an institution's financial condition. Interest rate risk relates to both sides of the balance sheet since the value of both assets and liabilities will change with changes in interest rates. The consensus is that the BIS recommendations will involve an additional capital requirement to cover excessive interest rate risk. This is also the view adopted by the FRB and OTS proposals. We now describe these proposals.

---

[10] It might appear that the model suggests that it is never optimal to invest in a bank. However recall that we are assuming equilibrium conditions here. If the bank has projects with a positive net present value these will be worthwhile investments and can attract fresh capital.

The FRB approach and the OTS approach both call for additional capital to support excessive levels of interest rate risk. It is assumed that banks and other deposit-taking financial institutions naturally assume a certain amount of interest rate risk in the normal course of business. The FRB and OTS approaches focus on identifying institutions that are outliers in terms of their interest rate risk exposure and requiring them to bolster their capital to support the excessive levels of interest rate risk. It is assumed that normal levels of interest rate risk are already covered by the current BIS risk-based capital requirements.

The FRB method uses a modified duration approach to approximate the net change in the economic value of each bank arising from a change in interest rates. The asset and liability cash flows are summarized in a compact matrix format and generic duration numbers are used to develop interest rate risk factors for each maturity band. The interest rate risk measure is a composite figure derived from the sum of weighted assets minus weighted liabilities plus an adjustment for off-balance sheet items. From these risk factors the dollar impact on the institution's net economic worth of a 100 basis points change in the level of interest rates is computed. The net interest rate exposure above a certain threshold is converted into a capital amount to cover interest rate risk.

The OTS approach to the measurement of interest rate risk is more complicated than the FRB approach although the essential thrust is the same. The OTS method uses the actual projected profile of the future asset and liability cash flows. To project these the cash flows the OTS method uses very specific and detailed assumptions on matters such as the prepayment of mortgage instruments and the interaction between prepayment levels and interest rate levels. Once the cash flows are projected the market value of the portfolio equity is computed under current market conditions and two additional interest rate scenarios corresponding to vertical shifts of 200 basis points in all yields. The fall in the market value of the portfolio equity under the most adverse of these shifts is computed as a percentage f of the bank's total assets. If f is less than 2 no additional capital is required. If f is greater than 2 the amount of additional capital to cover interest rate risk is

$$\frac{(f-2)V(0)}{2} \tag{10}$$

where $V(0)$ is the current asset value of the bank.

These two proposals focus directly on interest rate risk and provide no guidance on how – if at all – the additional capital required relates to that required for credit risk. It would be desirable to have a unified theoretical framework that could handle both types of risk. We discuss this in the next section.

4 Incorporation of Interest Rate Risk and Credit Risk

The option approach can be extend to incorporate both credit risk and stochastic interest rates. Rabinovitch (1989) has recently provided an extension of the Black Scholes model that incorporates stochastic interest rates. His approach uses the

Vasicek model to capture the stochastic evolution of interest rates. In this section we briefly discuss stochastic interest rate models and review the Vasicek model. We also present an option model which integrates both types of risk and discuss some of the policy implications of our analysis.

4.1 Stochastic Interest Rate Models

As a prelude to the analysis of interest rate risk we need to review how interest rates change over time. This leads to a discussion of stochastic interest rates. In the 1970's financial economists began to develop stochastic interest rate models inspired by the Black Scholes paper on the pricing of stock options. In these models the uncertainty was modelled through one or more so-called **state variables** which affect the prices of bonds and which evolve over time in a random fashion. These state variables have generally been modelled by continuous time stochastic processes. In the simpler models the short term interest rate process is taken to be the single state variable and such models are described as **one factor** models. By invoking certain assumptions the price of a generic bond is shown to be functionally related to the state variable. The same two key assumptions we met before underpin the derivation of these models: – the no arbitrage principle and the market completeness property. These models can be used to value options on bonds and other types of interest sensitive contingent claims.

The earliest model of this genre is due to Oldrich Vasicek (1977). Vasicek assumed that the short term rate of interest followed a mean-reverting process: i.e. that it wandered around some long run average level. The Vasicek model is the most tractable of the existing stochastic interest rate models but it permits the short term interest rate to become negative. This objectionable feature is absent in another popular model due to Cox Ingersoll and Ross (1985) (CIR). Both the Vasicek and CIR models are widely used since they both admit closed form solutions for bonds and options on bonds. However there are many other models of this type where the single state variable is the short term interest rate. A convenient summary of the major one factor stochastic interest rate models is given in a recent paper by Chan, Karolyai, Longstaff and Sanders (1992).

Academic researchers have also developed stochastic interest rate models based on two or more underlying factors. These first factor is often the short term interest rate and the second factor is normally either the long term interest rate or the inflation rate. Brennan and Schwartz (1979) proposed the earliest two factor interest rate models. Bond prices and option prices in this model are obtained by numerically solving a second order partial differential equations. Both Richard (1978) and Cox Ingersoll and Ross derived two factor models with closed form solutions for bond prices. Boyle (1980) and Chaplin (1987) extended the Vasicek model to incorporate two factors and presented closed form solutions for bond prices and in Chaplin's paper option on bonds as well. Jacobs and Jones (1986) have also developed a two factor stochastic interest rate model. The Jacobs and Jones model requires the numerical solution of a second order partial differential equation to evaluate bond prices and interest rate options.

In these two factor models parameter estimation poses significant econometric problems.

One deficiency of the stochastic interest rate models discussed so far is that they do not reproduce the current set of bond prices. We can see this intuitively if we note that these models are specified by a handful of parameters and that the current term structure has many more degrees of freedom than this. Dybvig (1989) among others illustrates how these models can be adjusted to overcome this problem. An alternative approach to ensuring that the interest rate model is compatible with the current set of bond prices was pioneered by Ho and Lee (1986). The Ho-Lee model takes as its starting point the existing set of bond prices and models interest rate uncertainty by a simple binomial lattice model in such a way so as to preclude arbitrage. Dybvig and others have pointed out some disadvantages of the Ho-Lee model. A more sophisticated model was developed by Heath Jarrow and Morton (HJM) (1992). The HJM model is designed to reproduce existing bond prices. HJM model uncertainty through the evolution of the forward rate process.

There are few empirical tests of these models and in particular there do not appear to be any comprehensive studies which undertake a comparative analysis of the validity of the different models. No doubt this is due in part to the numerical complexity of the solutions as well as the significant econometric problems involved in estimating the parameters of the models. Chan, Karolyai, Longstaff and Sanders (1992) have recently conducted empirical tests of the eight major one factor models based on US data for the period 1964-1989. Their central conclusion is that it is critical to model volatility correctly and that the best performing models are those which permit the interest rate volatility to depend on the level of the interest rate. They find that the two most popular one factor models the Vasicek model and the Cox Ingersoll and Ross square root process have poor performance relative to other less well known models such as Dothan (1980) and the Cox Ingersoll Ross (1985) "three halves" [11] model.

The published empirical investigations confirm the conventional wisdom that interest rate volatility is a key factor in the pricing of many types of interest sensitive claims. Part of the problem with many of the existing models is that they do not allow for changes in interest rate volatility. A recent model by Longstaff and Schwartz (1992) incorporates this feature. The Longstaff, Schwartz model does a much better job of explaining interest rate movements than the single factor models. They find that their model fits the term structure data well for maturities up to five years and has better explanatory power for longer maturity yields than competing models. Another advantage of their model is that closed form expressions are available for discount bonds and options on pure discount bonds. However the model involves six unknown parameters and their estimation poses challenging econometric problems.

---

[11] So called because the exponent of the short term rate which appears in the stochastic term is 3/2.

4.2 The Vasicek Model of Stochastic Interest Rates

In the Vasicek model the short term interest rate is assumed to follow an Ornstein Uhlenbeck mean reverting process. The changes in the short term rate arise from a drift term and a stochastic term. The drift term tends to pull the interest rate back towards its long term average value whereas the stochastic term consists of random fluctuations. We have

$$dr(t) = q[m - r(t)]dt + v\,dw(t)\,. \tag{11}$$

where

$r(t)$	is the short term interest rate
m	is the long run average value of r
v	is the standard deviation (volatility) of the interest rate process
q	is the speed of adjustment
$w(t)$	is a standard Weiner process

Vasicek derived and expression for the current price of a pure discount bond in this economy. He assumed that the so-called market price of risk λ was constant. The formula for the price at time t of a pure discount bond maturing at time T, is

$$P(r,t,T) = A(T - t)\exp[-rB(T - t)] \tag{12}$$

where

$$B(s) = \frac{1 - \exp(-qs)}{q}$$

$$A(s) = \exp\left[\bar{m}(B(s) - s) - \frac{(vB(s)/2)^2}{q}\right]$$

$$\bar{m} = m + \frac{v\lambda}{q} - \frac{1}{2}\left(\frac{v}{q}\right)^2\,.$$

The interest rate elasticity of the bond represents its sensitivity to changes in r standardized by dividing by the price of the bond. As noted by Boyle (1978) and Cox Ingersoll and Ross (1979) this is the stochastic analogue to the classical duration measure. The interest rate elasticity of the pure discount bond at time t with maturity at time T

$$\frac{1}{P}\frac{\partial P}{\partial r} = -B(T - t)\,. \tag{13}$$

The great advantage of the Vasicek model lies in its tractability. It suffers from the disadvantage that the interest rate can become negative.

4.3 Incorporation of Interest Rate Risk and Credit Risk

Let us assume that the market value, V of the bank's assets follows a geometric Brownian motion process

$$dV(t) = \mu V(t)dt + \sigma V(t)dz(t) \tag{14}$$

where μ and σ are constants.

This implies that the distribution of the asset returns after time T is lognormal with variance $\sigma^2 T$. We assume that the interest rate process follows the Vasicek model given by Equation (11). To explicitly model the relationship between the asset price volatility σ and interest rate volatility v assume that the z process and the w process are correlated with instantaneous correlation coefficient ρ. This means that movements in the interest rates affect the value of the bank's assets.

As before the equity of the banking firm is represented by a call option on the bank's assets. In this case we need to incorporate the stochastic interest rate. The formula derived by Rabinovitch corresponds to this situation. Hence the current value of the bank's equity is given by:

$$E(0) = V(0)N(D_1) - FP(r,0,T)N(D_2) \tag{15}$$

$$D_1 = \frac{\ln \dfrac{V(0)}{FP(r,0,T)} + \dfrac{\delta^2 \tau}{2}}{\delta\sqrt{\tau}}$$

$$D_2 = D_1 - \delta\sqrt{\tau}$$

$$\delta^2 \tau = \sigma^2 T + [T - 2B(T) + g(T)]\left(\frac{V}{q}\right)^2 - 2\rho\sigma(T - B(T))\frac{v}{q}$$

$$g(T) = \frac{1 - \exp[-2qT]}{2q}.$$

Note that if the interest rates are non-stochastic, v is zero and we recover the Merton (1977) result. The volatility parameter δ replaces the σ in Merton's formula. Note that δ includes terms arising from both interest rate risk and non-interest rate risk. In fact this formulation assumes that the value of the assets V changes as interest rates change. It is convenient to define the sensitivity of V to interest rate movements as

$$\Phi_V = \frac{1}{V}\frac{dV}{dr}. \tag{16}$$

Duan, Moreau and Sealy (1992) have derived an expression for the interest elasticity of the equity as follows

$$\Phi_E = \Omega[\Phi_V + B(T)] - B(T) \tag{17}$$

$$\Omega = N(D_1)\frac{V(0)}{E(0)}$$

The volatility of asset returns σ arises from two sources as follows:

Interest rate component of asset return volatility σ_{IR}
Non-interest rate component of asset return volatility σ_{NIR}

$$\sigma_{IR} = \rho\sigma$$
$$\sigma_{NIR} = \sqrt{(1-\rho^2)}\sigma$$
$$\sigma^2 = \sigma_{IR}^2 + \sigma_{NIR}^2$$

Furthermore the volatility of the bank's common stock can be expressed as:

$$\sigma_E = \sqrt{\Phi_E^2 v^2 + (\Omega\sigma_{NIR})^2} \tag{18}$$

This expression depicts the structural relationship between the volatility of the bank's equity, σ_E the elasticity of the equity with respect to interest rate risk, Φ_E the interest rate volatility, v and the non interest rate risk, σ_{NIR}. This relationship illustrates the interaction between credit risk and interest rate risk.

5 Summary

We have shown how the option-pricing model can be used to analyze current problems concerning risk-based capital for financial institutions. We started with a simple discrete-time derivation of the basic option formula that emphasizes the importance of the two basic economic intuitions of complete markets and no-arbitrage. Then we indicated how the continuous time Black Scholes model could be obtained.

The Black Scholes model provides a useful framework for the analysis of risk-based capital and the pricing of deposit insurance. We briefly reviewed the seminal contributions of Merton to this area. The option model illustrates the genesis of the moral hazard problem. The owners of a bank can increase their equity by increasing the risk of the assets. This increase is obtained at the expense of the deposit insurance fund. Different mechanism are available to curb this tendency. We discussed two of them: the imposition of additional capital requirements by the regulators and the existence of the bank's charter value.

At the present time the BIS has implemented specific capital adequacy guidelines to cover a bank's credit risk. The option model provides a theoretical framework for the discussion of this issue. The BIS is planning to extend these regulations to cover the additional capital required for interest rate risk. To derive a conceptual framework for the analysis of this issue we need a stochastic interest rate model. Although a number of models have been proposed none of them seems quite satisfactory. We drew attention to some recent work combining the Black Scholes model with the Vasicek model that represents a preliminary attack on this problem. This model provides a specific representation of the interaction between credit risk and interest rate risk. The model indicates that it is theoretically incorrect to consider these two risk separately and add together the capital requirements computed on an individual basis.

References

Arrow, K.: Le Role des valeurs boursieres pour la repartition la meillure des risques. *Econometrie*, Colloq. Internat. Centre National de la Recherche Scientifique 40 (Paris 1952), 41–47; discussion, 47–48 C.N.R.S. (Paris 1953).

Black, F., Scholes, M.: The Pricing of Options and Corporate Liabilites. *Journal of Political Economy*, **81** (1973) 637–654.

Boyle, P.P.: Immunization under Stochastic Models of the Term Structure. *Journal of the Institute of Actuaries*, **105** (1978) 177-188.

Boyle, P.P.: Recent Models of the Term Structure of Interest Rates with Actuarial Applications. *Transactions of the 21st International Congress of Actuaries*, (1980) 177-188.

Brennan, M., Schwartz, E.S.: A continuous Time Approach to the Pricing of Bonds. *Journal of Banking and Finance*, **3** (1979) 133–155.

Buser, S., Chen, A., Kane, E.: Federal Deposit Insurance, Regulatory Policy and Optimal Bank Capital. *Journal of Finance*, **36** (1981) 51–60.

Chan, K., C., Karolyi, G.A., Longstaff, F.A., Saunders, A.B.: An Empirical Comparison of Alternative Models of the Short-Term Interest Rate. *Journal of Finance*, **47, 3** (1992) 1209–27.

Chaplin, G.B.: The Term Structure of Interest Rates – A Model Based on Two Correlated Stochastic Processes with Closed Form Solutions for Bond and Option Prices. *Working Paper*, University of Waterloo, Waterloo, Ontario, Canada (1987).

Cox, J., Ingersoll, J., Ross, S.: Duration and the Measurement of basis Risk. *Journal of Business*, (1979).

Cox, J., Ross, S., Rubinstein, M.: Option Pricing: A Simplified Approach. *Journal of Financial Economics*. **7** (1979) 229–263.

On the Term Structure of Interest Rates. *Journal of Financial Economics*, **6** (1978) 59–69.

Duan, J-C., Moreau, A.F., Sealey, C.W.: Deposit Insurance and Interest Rate Risk: Pricing and Regulatory Implications. *Working Paper*, McGill University, Montreal (1992).

Duffie, D.: *Security Markets: Stochastic Models*, Boston: Academic Press (1988).

Dybvig, P.: Bond and Bond Option Pricing Based on the Current Term Structure. *Working Paper*, School of Business, Washington University, St Louis (1989).

FRB, A., Ronn, E.: A Characterization of the Daily and Intra-day Behavior off Returns on Options. *Working Paper*, Merrill Lynch Capital Markets and University of Texas at Austin. (1992).

Heath, D., Jarrow, R., Morton, A.: Bond pricing and the Term Structure of Interest Rates: A New Methodology for Contingent Claims valuation. *Econometrica*, **60, 1** (1992) 77–105.

Ho, T., Lee, S.: Term Structure Movements and Pricing Interest Rate Contingent Claims. *Journal of Finance*, **41** (1986) 1011–29.

Houpt, J.V., Embersit, J.A.: A Method for Evaluating Interest Rate Risk in US Commercial Banks. *Federal Reserve Bulletin*, (1991) 625–637.

Jacobs, R.L., Jones, R.A.: A Two Factor Latent Variable Model of the Term Structure of Interest Rates. *Working Paper*, Economics Department, Simon Fraser University, Burnaby, British Columbia, Canada (1986).

Kendall, Sarah B.: Bank Regulation under Nonbinding Capital Guidelines. *Journal of Financial Services Research*, **5** (1991) 275–286.

Longstaff, F., Schwartz, E.S.: Interest rate Volatility and the Term Structure: A two-factor General Equilibrium Model. *Journal of Finance*, **47** (1992) 1259–82.

Marcus, A.J.: Deregulation and bank Financial Policy. *Journal of Banking and Finance*, **8** (1984) 557–565.

Merton, R.C.: An Analytic Derivation of the Cost of Deposit Insurance and Loan Guarantees. *Journal of Banking and Finance*. **1** (1977) 3–11. California at Riverside.

Office of Thrift Supervision: Transmittal Number 63, September 11 1992, Office of Thrift Supervision, 1700 G Street NW., Washington DC 20552 (1992).

Rabinovitch, R.: Pricing Stock and Bond Options when the Default-Free rate is Stochastic. *Journal of Financial and Quantitative Analysis*, **24**, **4** (1989) 447–457.

Richard, S.: An Arbitrage Model of the Term Structure of Interest Rates. *Journal of Financial Economics*, **2** (1978) 33–57.

Vasicek, O.: An Equilibrium Characterization of the Term Structure. *Journal of Financial Economics*, **5** (1977) 177–188.

Immunization Theory: An Actuarial Perspective on Asset-Liability Management

Massimo De Felice

University of Rome "La Sapienza"

1 Introduction

Actuarial Traditions. The management of interest rate risk is a problem with deep roots in the actuarial literature. The results by Redington (1952) and by Haynes and Kirton (1952) may be considered path-breaking on a problem that, even then, "for many years must have intrigued so many actuaries"[1]. The purpose of their work is to analyse the financial structure of a life office and, in particular, the relationship between the assets and liabilities of a life assurance fund. The specific problem is how to determine the allocation of assets to make them, as far as possible, equally as vulnerable as the liabilities to those influences (typically the effects of fluctuations in the market rate of interest) which affect both. Redington adopts the word "immunization, to signify the investment of the assets in such a way that the existing business is immune to a general change in the rate of interest"[2]. Haynes and Kirton use the word "insulation" in a similar way. It is remarkable to consider how closely both the authors agreed in their fundamental conclusions. But it is also relevant, for critical purpose and for the sake of theory, to remark the difference in the approach to the problem: Haynes and Kirton paper "dealt primarily with matching, valuation being a by-product", while Redington's paper "dealt primarily with valuation, matching being a by-product"[3]. The first approach led to the development of strategies for cash-flow matching [4]. The second, which the present paper is concerned with, recognizes the central role of bond pricing theory and of models of the term structure of interest rates, which have undergone a strong development in the last twenty years.

Operational and Theoretical Problems. The original theoretical foundation by Redington appears relevant for the technical results (the Redington the-

---

[1] (Haynes and Kirton 1952, p. 198)

[2] (Redington 1952, p. 289)

[3] (Redington 1952, p. 316)

[4] For recent results see (Kocherlakota, Rosenbloom and Shiu 1988) and (Matrigali and Pacati 1993).

orem, in particular), and for the suggestions on financial management schemes and on the role to be attributed to capital markets. It is meaningful for firm theory that interest rate risk is considered one of the most important factor which can endanger the solvency of an office, and the view that assets and liabilities must be considered together. It is noteworthy for capital markets and financial institutions theory that attention has been paid to the foundations on which the investment service offered by a life office is based (Haynes and Kirton 1952, p. 141). Many of the problems with which immunization was concerned arise in a even more acute form in the case of private pension funds.

The original path-breaking papers contain a number of ideas relating to the theory of financial decision making (financial institutions management and selection of financial instruments) and to the operation of capital markets.

Starting from the 70's, interest rate risk has been considered a crucial problem in the management of banks and insurance companies, in highly market-oriented financial intermediation systems. This risk may have significant effects on the overall evolution of the financial system and it accounts for an important part of the work of the supervisory authorities (control of solvency). In the management of banking firms and of the financial section of insurance companies interest rate risk is an aspect of overall programmes (fund-raising and investment, coverage policies, use of equity capital) and is linked to maturity transformation. It is the major component of the so called "profit risk" and can be regarded, significantly, as a claim on net capital (in this sense intermediaries could be said to be self-insuring against interest rate fluctuations).

The original idea that in the control of interest rate risk "asset and liabilities must be considered together" has been incorporated in asset-liability management models (it appears evident in accordance with logic)[5]. This is important for its management relevance and is also a key to understanding the capital markets. The needs of borrowers and investors arising from their asset-liability management are considered one of the most important determinants of the demand for new financial instruments. It follows that "to comprehend the financial innovations that have occurred and are expected to occur in the future, a general understanding of the asset-liability management problem of major institutional investors is required"[6].

---

[5] The approach still suffers of organizational difficulties. Recently, Daykin and Hey (1990, p. 179) remarked that "assets and liabilities must be considered together. The traditional balance sheet approach does not encourage this. Assets and liabilities are on two different sides of the balance sheet; in many companies they may be seen as the responsibilities of quite different groups of people". Also Borch (1968, p. 331), referring to Redington, had emphasized the advantages of asset-liability management: "[...] investment and reinsurance decisions should be analyzed together and [...] the ultimate aim should be to find decisions which are jointly optimal. [...] Actuaries have discussed reinsurance arrangements for generations, often with the explicit purpose of reducing the fluctuations in the company's underwriting results, but have only occasionally indicated that these fluctuations may be cancelled, or accentuated by fluctuations in the company's investment results".

[6] (Fabozzi and Modigliani 1992, pp. XXI–XXII). Following the same line Bodie (1990,

Methodological Premise. It is only recently that interest rate risk has been defined and "measured" within a complete theoretical framework, taking uncertainty properly into account. Through the formulation of "semi-deterministic" and stochastic models of the term structure of interest rates, it has become possible to evaluate the so called interest rate sensitive (IRS) contracts (the value of which depends on the term structure of interest rates) in a way consistent with the random nature of the markets, and therefore to define interest rate risk in operational terms as the random effects produced on the contract's value by shifts in the market term structure.

From a methodological point of view, interest rate risk differs from that inherent in most risky securities. Interest rate sensitive contracts, in fact, usually have known cash flow profiles (amounts and timing of cash payments) and this gives to interest rate risk an "inherently intertemporal" dimension, related to investor's preferences on the optimal wealth allocation, which is difficult to specify if other securities (equities) are considered[7]. It follows that the typical schemes for the management of equity portfolios, based on the diversification rule, cannot be used in the case of interest rate sensitive portfolios. In the traditional Markowitz portfolio selection theory the security risk – risk of price fluctuations – is controlled by diversification considering a "sufficiently large" number of securities with imperfectly correlated returns – the less the correlation among security returns, the greater the impact of diversification on reducing variability. Things are quite different if we consider IRS portfolios, in which the value (price) of the contracts depends on the level of interest rates. In this case, the risks are highly correlated, hence the risk of investment portfolios cannot be reduced by diversification. In the case of intermediation portfolios (with assets and liabilities), interest rate risk can be controlled by selecting contracts as "similar" as possible (in the sense of immunization) with opposite account sign (perfect matching is the extreme case, with perfectly negative correlation). The asset-liability management models then represent a simple but relevant way to profit from negatively correlated opportunities.

2 Basic Results in Immunization Theory

The typical IRS portfolio may be characterized by considering its asset and liability streams. These streams will consist, in general, of random payments which are function of the interest rate.

---

p. 43) notices that "while there is no comprehensive or universally accepted theory of financial intermediaries in general, or of insurance companies in particular, most scholars believe that the key to understanding the funding and investment policies of these institutions is the matching of assets and liabilities" in the sense of immunization. See also (Bodie 1989, p. 28).

[7] Interest rate risk is "inherently intertemporal" in the sense that "an unexpected change in the interest rate now affects all future returns, so interest rate risk compounds over time. On the other hand, security risk is the contemporaneous resolution of returns" (Ingersoll 1987, p. 404).

Let c_A and c_B represent, respectively, the contractual parameter vectors of assets and liabilities and let us denote by $P(t; c_A)$ and $P(t; c_B)$ their values at time t: the solvency condition requires the net present value $P_N(t) = P(t; c_A) - P(t; c_B)$ to be non-negative. Since the future value $P_N(t')$, $t' > t$, is in general unknown at time t, the portfolio will be a risky portfolio.

Financial immunization theory can be defined as the theory for controlling $P_N(t)$ uncertainty, i.e. the theory for controlling portfolio solvency.

In the original Redington formulation, price is the operational variable for the control of financial risk, and the construction of the control plan is based on the construction of a pricing model for marketed financial contracts. In what follows, we will refer to stochastic immunization or to semi-deterministic immunization to indicate that the financial control strategy is respectively stated with or without considering the probability distribution of the relevant random variables.

2.1 The Semi-Deterministic Approach to Immunization

Assumption on the Markets. In this framework, the market structure at time t can be identified by the structure of prices, $v(t, s)$, of the unit zero coupon bonds (ZCB), that is the bonds which pay with certainty one unit of money at time $s \geq t$. As a function of s, $v(t, s)$ are the discount factors for the time periods $[t, s]$. The price structure is also described by the force of interest, defined as $\delta(t, s) = -\frac{\partial}{\partial s} log v(t, s)$.

The classical approach to immunization theory (used by Redington) is based on the so-called "additive shift hypothesis", which amounts to assuming:

$$\delta(t', s) = \delta(t, s) + Z(t, t') , \quad t \leq t' \leq s , \tag{1}$$

where Z is a random variable, independent of s, representing the size of the additive shift which the yield curve undergoes in the time interval between t and t' (since Z is independent of s, disturbances between t and t' do not change the shape of curve $\delta(t, s)$). Assumption (1) can be significantly generalized to the so-called "arbitrary shift hypothesis", that is:

$$\delta(t', s) = \delta(t, s) + Z(t, t', s) , \quad t \leq t' \leq s ; \tag{2}$$

the fact that Z also depends on s implies that a single shift can alter the form of the yield curve arbitrarily.

Without loss of generality we shall consider additive shifts having effect at time t^+, immediately after t ($t^+ = t + dt$). Therefore, (2) reduces to

$$\delta(t^+, s) = \delta(t, s) + Y(s) , \quad s \geq t^+ ,$$

where $Y(s)$ is the shift occurred in t^+. If the random function $Y(s)$ is differentiable and if

$$Y^2(s) \geq Y'(s) , \quad s \geq t^+ ,$$

then the shift $Y(s)$ is called a "convex shift". Under the additive shift assumption

$$\delta(t^+, s) = \delta(t, s) + Y , \quad s \geq t^+ ,$$

with Y constant (any additive shift is convex). The probability distribution of the random variable Z (or Y) is not considered. Therefore the immunization theory based on the arbitrary shift hypothesis can be defined as "semi-deterministic"[8].

Financial Indices of Payment Streams. The semi-deterministic immunization theory applies to financial contracts with cash flows known at the evaluation time[9]. The classical approach, based on the additive shift hypothesis, is a special case of a semi-deterministic theory. The contractual features of any contract define, at time t, a vector x which represents a stream of non-negative payments $x_1, x_2, \ldots, x_m$, due at the dates $t_1, t_2, \ldots, t_m$ $(t \le t_1 \le t_2 \le \ldots \le t_m)$.

With respect to the vector x, the following financial indices can be defined:

$$P(t; x) = \sum_{k=1}^{m} x_k v(t, t_k) , \tag{3}$$

$$D(t; x) = \sum_{k=1}^{m} (t_k - t) p_k , \tag{4}$$

$$MAD_j(t; x) = \sum_{k=1}^{m} |t_j - t_k| \, p_k , \tag{5}$$

$$M^{(2)}(t; x) = \sum_{k=1}^{m} [t_k - D(0; x)]^2 p_k , \tag{6}$$

where $p_k = x_k / \sum_{k=1}^{m} x_k v(t, t_k)$.

The price P is defined in (3) as a linear functional, definition (4) is the Macaulay duration (1st order moment of the normalized time distribution of the discounted cash-flows), definition (5) is the mean absolute deviation (MAD) of the time distribution with respect to a given payment date t_j, $M^{(2)}$ is the time-spread (the 2nd order central moment of the time distribution). The standard time-spread is defined as the square root of $M^{(2)}(t, x)$.

The typical situation of a financial intermediary is represented by the aggregate asset vector $x = \{x_1, x_2, \ldots, x_m\}$ and the aggregate liability vector $y = \{y_1, y_2, \ldots, y_m\}$, both having non-negative cash flows on the same time grid $\{t_1, t_2, \ldots, t_m\}$. We define the margin stream $z = x - y$, the net present value $P_N(t) = P(t; x) - P(t; y)$ (obviously it results $P_N(t) = P(t; z)$), and the net time spread of the portfolio $M_N^{(2)}(t) = M^{(2)}(t; x) - M^{(2)}(t; y)$.

[8] (De Felice and Moriconi 1991b)
[9] Semi-deterministic immunization schemes may also be applied to floating rate notes with "synchronous indexation", see (De Felice, Moriconi and Salvemini 1993, p. 134), (Moriconi 1991).

The Classical Immunization Theorem.

Theorem 1. *Let us consider, at time t, the term structure defined by the function* $\delta(t,s)$, $\forall\, s \geq t$, *the liability stream* $\boldsymbol{y} = \{y_1, y_2, \ldots, y_m\}$ *and the asset stream* $\boldsymbol{x} = \{x_1, x_2, \ldots, x_m\}$.
Assume that

$$P(t; \boldsymbol{x}) = P(t; \boldsymbol{y}) \ . \tag{7}$$

Under the assumption:

$$\delta(t^+, s) = \delta(t, s) + Y \ , \quad s \geq t^+ \ ,$$

it will be

$$P(t^+; \boldsymbol{x}) \geq P(t^+; \boldsymbol{y}) \tag{8}$$

if and only if

$$D(t; \boldsymbol{x}) = D(t; \boldsymbol{y}) \ ,$$
$$MAD_j(t; \boldsymbol{x}) \geq MAD_j(t; \boldsymbol{y}) \ , \quad j = 1, 2, \ldots, m \ . \tag{9}$$

$\square$

The theorem holds, in general, for convex shifts. It generalizes the Redington model (based on the assumption of infinitesimal additive shifts) and the Fisher and Weil model (which allows the immunization of a single liability)[10]. The classical condition of financial immunization (8) ensures that there is a possibility of making a profit "whatever happens": it is for this very reason that the additive shift assumption is not consistent with market equilibrium[11].

Criticism of the additive shift hypothesis (which already began alongside Redington's original work[12]) has not prompted a revision of the foundations of the method. A "more accurate description"[13] of the shift process is needed, but the response has been to blame the poor empirical results of the model rather than the overly naïve approach to uncertainty.

By adopting alternative assumptions on the shifts (multiplicative, additive-multiplicative) it has been possible to define ad-hoc immunization procedures

---

[10] (Fisher and Weil 1971). In the case of a single liability the MAD constraints are, in fact, automatically satisfied.

[11] According to Theorem 1 it could be possible to create a portfolio that "leaves an investor better off than holding a pure discount bond under some conditions and no worse off under all conditions" (Ingersoll, Skelton and Weil 1978, p. 636): in the simplest form, one could build a "money pump" by constructing a portfolio containing two zero-coupon bonds financed by selling a zero-coupon bond with the same time to maturity as the duration of the portfolio.

[12] In the discussion of Redington (1952) paper, Rich (p. 319) remarked "how delightful it would be if the funds of a life office could be so invested that, on any change in the rate of interest – whether up or down – a profit would always emerge!". This consideration is referred to in the literature as the "Rich paradox".

[13] (Fisher and Weil 1971, p. 417)

(each one relevant to the particular hypothesis on the shift)[14] which do not go beyond the theoretical limits and possibilities of application of the classical approach. In any case, the argument that an ad-hoc hypothesis gives a better empirical result than others seems specious since "unfortunately actual rate movements are far more varied than the theoretical categories that have been proved immunizable"[15] according to (8).

The Downside Risk Theorem. None of the ad-hoc hypotheses on the random evolution of the yield curve may be considered empirically more realistic than the other. It is then relevant to find the strategy that is "least sensitive" to the specific assumption on the shift. If arbitrary shifts are considered (having effects at t^+), we can state the following:

Theorem 2. *Let us consider, at time t, the term structure defined by the function $\delta(t, s)$, $\forall s \geq t$, the liability stream: $y = \{y_1, y_2, \ldots, y_m\}$ and the asset stream $x = \{x_1, x_2, \ldots, x_m\}$.*
  Assume that

$$P(t; x) = P(t; y) \ . \tag{10}$$

Under the assumption

$$\delta(t^+, s) = \delta(t, s) + Y(s) \ , \quad s \geq t^+ \ , \tag{11}$$

being $Y(s)$ a function with continuous, upper-bounded derivative, and if the following conditions hold

$$D(t; x) = D(t; y) \ , \tag{12}$$
$$MAD_j(t; x) \geq MAD_j(t; y) \ , \quad j = 1, 2, \ldots, m \ , \tag{13}$$

then

$$P(t^+; z) \geq P(t; y)KM_N^{(2)}(t) \ , \tag{14}$$

where K is a real valued random variable, independent of x and y, characterized only by the form of the shift. □

Theorem 2 requires streams x and y to meet the same conditions as Theorem 1, but in general does not guarantee the immunization condition $P(t^+; z) \geq 0$. Under hypothesis (11), conditions (10), (12) and (13) provide a lower bound on the post shift value of the net stream. The lower bound is expressed as the product of a quantity $P(t; y)M_N^{(2)}(t)$, which is characteristic of the streams at time t (and therefore known at t), and of a factor K, a real-valued random variable. Thus the net post-shift value $P(t^+; z)$ is also a random variable which may have negative, nil or positive realizations. It is clear from (14) that the closer to zero is $M_N^{(2)}(t)$ the more limited will be the effect of an "unfavorable" change in the term structure (i.e. a negative value of K) on the financial equilibrium

---

[14] See (Bierwag, Kaufman and Toevs 1983, pp. 145–147), (Bierwag 1987, pp. 258–286).
[15] (Granito 1984, p. 38)

between x and y after the shift. In this sense $M_N^{(2)}(t)$ can be used as a measure of the exposure of the portfolio to any interest rate fluctuations.

It is superfluous but nevertheless illustrative to note that portfolios immunized in the classical sense are not exposed to risk induced by additive shifts in the yield structure. It is therefore justifiable to interpret $M_N^{(2)}(t)$ as a measure of risk (immunization risk) for portfolios that are only immunized against convex shifts (in particular against additive shifts). The measure of risk is not derived from assessments on the future state of the market, but from structural inequalities (as such, it measures the downside risk). As $M_N^{(2)}(t)$ is known at time t, it can be used as a strategic variable in management procedures.

The variable K is given by $\inf_{s \geq t} f''(s) = K$, where f is the shock function: $f(s) = v(t^+, s)/v(t, s)$, $s \geq t$. It can be shown that if K is assumed to have non-negative values (as in the case of additive shifts) (14) will give the classical immunization condition $P(t^+; z) \geq 0$ as stated in Theorem 1. From Theorem 2 it follows that portfolio selection and management methods based on the maximization of the time-spread (of which "maximum convexity" strategies[16] are a special case) are not reliable operationally; they are only suitable for a market in which the yield structure is subject to additive shifts.

It is important to note that a portfolio which is immunized at time t (according to Theorem 1) remains immunized as time passes if the basic conditions for immunization do not change, i.e. if the yield curve evolves in a purely deterministic manner and if cash-flows in the payment streams x and y do not mature (Conservation Theorem). Theorems 1 and 2, and the Conservation Theorem define the theoretical framework of semi-deterministic immunization, and "rearrange" a set of results whose original nucleus can be attributed to Fong and Vasicek[17]. The "minimum risk" approach represents the final step in the semi-deterministic theory of financial immunization: more general pricing models and interest rate risk control strategies can only be defined if the market model is broadened to take into account economic agents assessments about the future, which are formulated as probability evaluations on states of nature.

2.2 The Stochastic Immunization Theory

The Market Model. Single factor diffusion models for the pricing of default-free bonds are typically based on the following main assumptions:

a) the single variable that determines the state of the economy at time t is the spot rate of interest $r(t)$, that is, the yield on an instantaneously maturing ZCB;

b) the spot rate follows a diffusion process described by the Ito stochastic differential equation:

$$dr(t) = f(r, t)\, dt + g(r, t)\, dZ(t) \ , \tag{15}$$

---

[16] See (Kopprash 1985), (Klotz 1985), (Fabozzi and Modigliani 1992, p. 401).

[17] (Fong and Vasicek 1982), (Fong and Vasicek 1984). For the proofs of the theorems see (De Felice and Moriconi 1991b, Chap. 3); see also (Shiu 1986), (Shiu 1987), (Montrucchio and Peccati 1991).

where f and g^2 are the instantaneous drift and variance, respectively, and Z is a standard Brownian motion;

c) the market is perfect and frictionless: trading in all assets takes place continuously, there are no taxes or transaction costs, riskless arbitrage opportunities are precluded, the investors have "homogeneous expectations" [18].

Under these assumptions, the model price $P(t)$ of any IRS asset traded on the market at time t, as well as the price of any portfolio of such bonds (contracts), can be modeled as a function of the state variable $r(t)$, of time, and of a vector c of contractual parameters: $P(t) = P(r,t;c)$. Applying Ito's lemma and using a no-arbitrage argument, we obtain the general valuation equation:

$$P_t + f^* P_r + \frac{1}{2} g^2 P_{rr} - rP = 0 \ , \tag{16}$$

where $f^*(r,t) = f(r,t) + q(r,t)g(r,t)$ is the *risk-adjusted* drift and $q(r,t)$ is a function, independent of c, which represents the *market price of risk*.

The partial differential equation (16), together with appropriate boundary conditions, can be used to price any IRS financial claim. In particular, the price $v(t,s)$ of a unit ZCB maturing at time s can be obtained by solving (16) under the condition $v(s,s) = 1$. The function $v(t,s)$ represents the equilibrium discount factors and determines the term structure of interest rates prevailing on the market at time t as a function of maturity, s.

In these models a natural measure of basis risk for any IRS financial claim is given by the semielasticity of the price with respect to the state variable:

$$\Omega(r,t;c) = -\frac{P_r(r,t;c)}{P(r,t;c)} \ .$$

In many cases, it is possible to express the basis risk in units of time by defining the stochastic duration $D(r,t;c)$ for the claim c as the time-to-maturity of a ZCB with the same risk, that is:

$$D(r,t;c) = \varphi^{-1}\{\Omega(r,t;c)\} - t \ ,$$

where $\varphi^{-1}(\cdot)$ is the inverse function, with respect to s, of $\varphi(r,t,s) = -\frac{v_r(t,s)}{v(t,s)}$.

The Stochastic Immunization Theorem. Let c_A and c_L be the contractual vectors representing the portfolios of assets and liabilities, respectively, of a financial intermediary.

Theorem 3. *Let the asset portfolio have the same price of the liability portfolio at time t:*

$$P(r,t;c_A) = P(r,t;c_L) \ . \tag{17}$$

If

$$\Omega(r,t;c_A) = \Omega(r,t;c_L) \ , \tag{18}$$

---

[18] See e.g.(Cox, Ingersoll and Ross 1985).

then

$$dP(r, t; c_A) = dP(r, t; c_L) \; . \tag{19}$$

$\square$

Condition (18) can usually be stated in the form of the so-called "duration constraint":

$$D(r, t; c_A) = D(r, t; c_L) \; .$$

The stochastic immunization theorem is derived within a no-arbitrage framework[19] and is consistent with market equilibrium. In this sense, the definition of stochastic immunization is a way to consider Redington's point of view, and allows to remove Rich "paradox"[20].

If the portfolios c_A and c_L can be selected fulfilling conditions (17) and (18), the financial position of the intermediary is instantaneously riskless and the asset-liability portfolio is said to be stochastically immunized. Since the present value and the duration constraints are defined at time t, the stochastic immunization theorem is the methodological reference to construct portfolio selection procedures.

The stochastic immunization theorem allows to select portfolios which are instantaneously riskless at time t. Since the spot rate $r(t)$ varies continuously and randomly, the present value and the duration constraints will not be fulfilled immediately after time t. So, for its theoretical basis, stochastic immunization (as defined with reference to diffusion models) requires continuous adjustments of portfolio compositions[21].

The Normalized Prices. In the framework of the stochastic immunization theory the only relevant features of any bond are represented by its price and by its risk measure. In particular, any bond, or portfolio of bonds, can be considered equivalent to a ZCB having the same price and the same duration. By adopting this point of view it is possible to develop a unified approach for the valuation of all the bonds traded on the same date on the actual market and to provide a market-oriented interpretation of stochastic asset-liability management[22].

Let us consider the default-free bonds traded on the actual market at time t. By applying the previous theory, we obtain an estimate of the current term structure $v(t, s)$. Moreover, for any bond traded on the market, the model will provide the theoretical price $P(t)$ and the relative stochastic duration $D(t)$. Typically, the price $q(t)$ observed on the market will differ from the "equilibrium" price $P(t)$; therefore, the model can be used to identify underpriced and overpriced bonds. However, the price is not the only relevant characteristic of the bond and in order to derive a consistent valuation of mispricing we have also to

---

[19] (Castellani, De Felice and Moriconi 1992a)

[20] (Boyle 1978, p. 180)

[21] An example of application of the stochastic immunization theorem to discrete time models can be found in (Mari 1992).

[22] This approach is defined in (Castellani, De Felice and Moriconi 1992b).

take into account the risk measure. Given a bond c with model price $P(t; c)$ and duration $D(t; c)$, let us define the *ex-ante face value* of the bond as:

$$z(t, t + D) = \frac{P(t; c)}{v(t, t + D)} \ ;$$

this quantity represents the face value of the ZCB with time-to-maturity $D = D(t; c)$ which has the same model price at time t as the bond c. Obviously, for a unit ZCB with maturity s it must be $z(t, s) = 1$.

We can now define the *normalized price* of the bond c with market price $q(t)$ as:

$$\Pi(t; c) = \frac{q(t)}{z(t, t + D)} = \frac{q(t)}{P(t; c)} v(t, t + D) \ .$$

The normalized price can be interpreted as the market discount factor for the time horizon D. More precisely, it is the price prevailing in the market at time t of a bond equivalent to a unit ZCB maturing after $D(t; c)$ time periods. This market discount factor must be compared with the model discount factor $v(t, t + D)$, that is with the time t model price of the unit ZCB maturing at time $t + D$.

With the above definition any bond (or, in general, IRS contract) traded on the market is identified by the pair (D, Π) and can be graphically represented in this plane. If $q(t) = P(t; c)$, then $\Pi(t; c) = v(t, t + D)$. Hence, we interpret the graph of the function $v(D) = v(t, t + D)$ in the plane (D, Π) as the *market equilibrium line* prevailing at time t. Bonds which are considered underpriced, that is with $q(t) < P(t; c)$, will be represented by points lying below the equilibrium line. Overpriced bonds will plot above this line[23].

3 Asset-Liability Management Models

Structure of IRS Portfolios. An outstanding intermediation portfolio can be formally represented at the evaluation time t by the matrix $A = \{a_1, a_2, \ldots, a_n\}$ of the n asset contractual vectors, and by the matrix $B = \{b_1, b_2, \ldots, b_n\}$ of the n liability contractual vectors[24]. The time structure of the intermediation portfolio is referred to the time period $[t, t_m]$; in particular, it is defined by

---

[23] The "vertical" displacement from the equilibrium line of the normalized price of c can be expressed as:

$$\Delta(t; c) = \Pi(t; c) - v(t, t + D) = \varepsilon(t; c) v(t, t + D) \ ,$$

where:

$$\varepsilon(t; c) = \frac{q(t)}{P(t; c)} - 1$$

is the relative pricing error. Hence, the mispricing in the (D, Π) plane, as measured by $\Delta(t; c)$, is the relative pricing error discounted over a time horizon equal to the bond duration.

[24] The fact of considering the same number of assets and liabilities does not involve any loss of generality.

the time vector $t = \{t_1, t_2, \ldots, t_m\}$. If the portfolio consists of contracts with deterministic cash-flows, A is a $(n \times m)$ matrix of non-negative elements, where the generic element a_{ik} represents the amount to be received at time t_k for the i-th asset. Similarly, B is a $(n \times m)$ matrix whose generic element b_{jh} represents the amount to be paid at time t_h for the j-th liability.

Let $q^{(a)} = \left\{ q_1^{(a)}, q_2^{(a)}, \ldots, q_n^{(a)} \right\}$ and $q^{(b)} = \left\{ q_1^{(b)}, q_2^{(b)}, \ldots, q_n^{(b)} \right\}$ represent, at time t, the vectors of the market prices of the assets and the liabilities, respectively[25].

The vectors $\alpha = (\alpha_1, \alpha_2, \ldots, \alpha_n)$ and $\beta = (\beta_1, \beta_2, \ldots, \beta_n)$ give, respectively, the shares of the contracts of basket A and of basket B held in the asset and liability portfolios. If contracts involving deterministic payments are considered, the asset portfolio will have cash flow x, with individual components due at time t_k given by $x_k = \sum_{i=1}^{n} \alpha_i a_{ik}$, $k = 1, 2, \ldots, m$; the cash flow y of the liability portfolio will be defined in a similar way. Let C be the cost function and C_a and C_b the volume (cost) of the assets and the liabilities. In portfolio selection problems, the pairs $\left(\overline{\alpha}_i, \overline{\overline{\alpha}}_i \right)$ and $\left(\overline{\beta}_i, \overline{\overline{\beta}}_i \right)$, $i = 1, 2, \ldots, n$, can be used to indicate the asset and liability portfolios composition constraints, assumed to be non-negative (i.e. ruling out short selling), where we may have $\overline{\overline{\alpha}}_i = +\infty$, $\overline{\overline{\beta}}_i = +\infty$.

This scheme can be used to select and manage an investment portfolio (a bonds portfolio, for example) dedicated to cover a fixed liability stream. In the case of a single liability, $L > 0$, to be paid at time H $(t < H < t_m)$ the variable L assumes the strategic role of a "target" variable (planned value at the end of the holding period).

3.1 The Semi-Deterministic Approach to Asset-Liability Management.

In line with the original spirit of immunization[26] the general scheme for a semi-deterministic model of asset-liability management can be stated with reference to the following:

Problem. Given the portfolio composition constraints and baskets A and B of assets and liabilities, select the "optimally immunized" intermediation portfolio (α, β), that is the portfolio which is immunized against convex shifts and carries minimum risk with respect to arbitrary shifts.

---

[25] In the case of many financial instruments we can properly speak of quotations if market prices are quoted. Usually, and especially in the case of loans or policies, we can consider strategic prices, based on management criteria: for example, by calculating the present value of the payment stream (using a suitable term structure of interest rate), or by taking the outstanding principal of the respective redemptions, or by considering the mathematical reserve.

[26] Significantly, Redington (1952, p. 292) points out that immunization is "against profit as well as loss".

Referring to Theorem 2, the problem can be written as:

$$\min \left\{ \sum_{i=1}^{n} \alpha_i \sum_{k=1}^{m} t_k^2 a_{ik} v(t, t_k) - \sum_{i=1}^{n} \beta_i \sum_{k=1}^{m} t_k^2 b_{ik} v(t, t_k) \right\} \tag{20}$$

$$\text{s.t.} \quad \sum_{i=1}^{n} \alpha_i \sum_{k=1}^{m} a_{ik} v(t, t_k) - \sum_{i=1}^{n} \beta_i \sum_{k=1}^{m} b_{ik} v(t, t_k) = 0 , \tag{21}$$

$$\sum_{i=1}^{n} \alpha_i \sum_{k=1}^{m} t_k a_{ik} v(t, t_k) - \sum_{i=1}^{n} \beta_i \sum_{k=1}^{m} t_k b_{ik} v(t, t_k) = 0 , \tag{22}$$

$$\sum_{i=1}^{n} \alpha_i \sum_{k=1}^{m} |t_j - t_k| a_{ik} v(t, t_k) - \sum_{i=1}^{n} \beta_i \sum_{k=1}^{m} |t_j - t_k| b_{ik} v(t, t_k) \geq 0 , \tag{23}$$

$$j = 1, 2, \ldots, m ,$$

$$\sum_{i=1}^{n} \alpha_i q_i^{(a)} = C_a , \tag{24}$$

$$\sum_{i=1}^{n} \beta_i q_i^{(b)} = C_b , \tag{25}$$

$$\overline{\alpha}_i \leq \alpha_i \leq \overline{\overline{\alpha}}_i , \quad i = 1, 2, \ldots, n , \tag{26}$$

$$\underline{\beta}_i \leq \beta_i \leq \overline{\overline{\beta}}_i , \quad i = 1, 2, \ldots, n . \tag{27}$$

Expressions (21) and (22) require the selected streams to meet the present value constraint and the duration constraint, the m conditions (23) formalize the MAD constraints, (24) and (25) require the asset and liability volumes to be at the levels C_a and C_b respectively; the conditions (26) and (27) define the set of possible values of the weights of individual assets in the portfolio composition.

It is worth noting that C_a and C_b are strategic inputs, exogenous to the problem. Specifically, if $q^{(a)}$ and $q^{(b)}$ were equilibrium prices with respect to the term structure $v(t, s)$, then $C_a = P(t; x)$ and $C_b = P(t; y)$ (if $C_a = C_b$ we have the so called "full utilization" situation).

The scheme may be used for the selection of new investment portfolios as well as for managing outstanding ones (e.g. bond portfolios) in order to cover a fixed liability stream, that is, β.(or, equivalently, y) being fixed. In this case, it is meaningful to consider both the minimum risk and the minimum cost portfolio. It is then possible to determine, in the (cost, risk) plane, the efficient frontier representing the so called risk/return trade-off as the locus of coverage portfolios carrying minimum risk for any fixed level of cost. Referring to a given liability stream, the extreme points of the empirical efficient frontier are the portfolios derived as "solutions" of the problems:

$$\min \sum_{i=1}^{n} \alpha_i q_i^{(a)} \tag{28}$$

and

$$\min \sum_{i=1}^{n} \alpha_i \sum_{k=1}^{m} (t_k - t)^2 a_{ik} v(t, t_k) \; , \qquad (29)$$

under the present value, the duration, the MAD and the portfolio composition constraints.

After the extreme points of the efficient frontier have been calculated, the range for the cost of the possible coverage portfolios, I, is identified. Other points of the frontier will be obtained as the solution to problem (29) for fixed cost levels, i.e. considering the new constraint:

$$\min \sum_{i=1}^{n} \alpha_i q_i^{(a)} = c \; ,$$

where $c \in I$.

A consideration of operational relevance is that, in the case of a single liability, the MAD constraints are automatically satisfied. In this case, if x is the portfolio selected in order to cover the liability L at time H ($L > 0$, $t < H < t_m$), it will have the "immunized" return $[L/P(t; x)] - 1$ in the holding period $H - t$. Time H can be interpreted as the "optimal investment time". This idea is a generalization, in the context of "minimum risk immunization", of the original concept proposed by Fisher and Weil (1971, p.416).

3.2 The Stochastic Approach to Asset-Liability Management.

In the stochastic framework also it is natural to define strategies for asset liability management which minimize the cost of the net portfolio. Typically, this optimization problem assumes a linear form with objective function:

$$\min \left[\sum_{i=1}^{n} \alpha_i q_i^{(a)} - \sum_{i=1}^{n} \beta_i q_i^{(b)} \right] \; , \qquad (30)$$

with constraints on present value $P(r(t), t; \alpha) = P(r(t), t; \beta)$, and on riskiness $\Omega(r(t), t; \alpha) = \Omega(r(t), t; \beta)$, to which we must add a volume constraint $\sum_{i=1}^{n} \alpha_i q_i^{(a)} = C_a$, or $\sum_{i=1}^{n} \beta_i q_i^{(b)} = C_b$, as well as the usual composition constraints $\overline{\alpha}_i \leq \alpha_i \leq \overline{\overline{\alpha}}_i$, $\overline{\beta}_i \leq \beta_i \leq \overline{\overline{\beta}}_i$, $i = 1, 2, \ldots, n$.

Alternative objective functions can be defined in the practical implementation of the theory. As we previously mentioned, fulfilling at time t conditions (18) and (19) in Theorem 3 only guarantees that the asset-liability portfolio is instantaneously riskless on that date. Since portfolio rebalancing can be performed only at discrete times, actual immunization strategies are only an approximation to the idealized immunization achieved by continuous rebalancing. In many applications the rebalancing period $\tau > 0$ is fixed and known at time t. It is then possible to define as objective function the expectation on some portfolio characteristic at the next rebalancing time, $T = t + \tau$.

For the sake of simplicity, let us assume that $\tau < t_1 - t$, that is the next portfolio adjustment is performed before the first payment date. The fact that the present value constraint is satisfied at time t does not imply, in general, that $E_t[P_N(r(t+\tau), t+\tau)] = 0$. Therefore it is possible to apply immunization trying to maximize this expectation. So we are led to the following:

Problem. Select at time t from basket A and from basket B the immunized intermediation portfolio (α, β) with maximum expected value, given the portfolio composition constraints and the volume constraints.

This corresponds to the linear program:

$$\max \ E_t\left[P_N(r(T), T)\right] , \tag{31}$$
$$\text{s.t.} \ P(r(t), t; \alpha) = P(r(t), t; \beta) ,$$
$$\Omega(r(t), t; \alpha) = \Omega(r(t), t; \beta) ,$$
$$\sum_{i=1}^{n} \alpha_i q_i^{(a)}, = C_a ,$$
$$\sum_{i=1}^{n} \beta_i q_i^{(b)}, = C_b ,$$
$$\underline{\alpha}_i \leq \alpha_i \leq \overline{\overline{\alpha}}_i , \quad i = 1, 2, \ldots, n ,$$
$$\underline{\beta}_i \leq \beta_i \leq \overline{\overline{\beta}}_i , \quad i = 1, 2, \ldots, n .$$

Stochastic immunization strategies oriented toward perfect matching can be also considered by introducing minimum risk objectives. It seems relevant to minimize the probability, evaluated at time t, that the net present value of the portfolio at time T will be greater (in absolute value) than a level a, strategically fixed at time t. An upper bound for this probability is immediately given by Chebychev's inequality:

$$\mathbf{P}\left[|P_N(r(t+\tau), t+\tau)| \geq a\right] \leq \frac{E_t\left[P_N^2(r(t+\tau), t+\tau)\right]}{a^2} .$$

It is then consistent to consider the following:

Problem. Select from basket A and from basket B the immunized intermediation portfolio (α, β) carrying minimum risk (i.e. minimum 2-nd order moment of the random variable $P_N(T)$), given the portfolio composition constraints and the volume constraints.

This selection problem, obtained replacing the objective (31) by:

$$\min E_t\left[P_N^2(r(T), T)\right] \tag{32}$$

is a quadratic programming problem.

In order to define both the previous problems we must first choose a stochastic model for the bond market and specify the contractual details of the securities in baskets **A** and **B**[27].

Obviously, the problems defined by the objective functions (31) and (32) can also be solved in the case of contracts with deterministic cash flows[28].

In the problem having the objective $\max \mathbf{E}_t [P_N(T)]$, the selected portfolio will be the one which, among all the instantaneously riskless portfolios, maximizes the expectation (at the evaluation time t) of the net value at the fixed future time $T = t + \tau$. This is the same as if the portfolio were selected to be instantaneously riskless at time t and such that it would emerge "in the best condition as possible" (in the sense of expected value) at the next market opening[29]. Obviously, the maximum value objective does not allow for control on the future level of risk.

The procedure with the objective function $\min \mathbf{E}_t \left[P_N^2(T) \right]$ instead minimizes the probability (evaluated at time t) that the net absolute value of the portfolio (and therefore the present value mismatching) at the next fixed rebalancing time T will be greater than a "threshold" given at time t.

The two procedures allow to identify the efficient frontier of the possible intermediation portfolios: this leads to a redefinition the traditional "mean-variance" approach in terms of stochastic immunization. Obviously, also in the stochastic framework, procedures for the selection of an immunized investment portfolio covering a fixed liability stream are immediately obtained as special cases of the previous asset liability management problems.

4 Applications and Examples

Surplus Stream. The concept of surplus stream was defined to answer the management need of identifying the time allocation of surplus. The idea has been originally introduced in the semi-deterministic framework[30]. Given an outstanding intermediation portfolio, the programming problem (20) can be solved by constraining the composition to be equal to 1 for the liability stream ($\beta_i = 1$ for each i), less than or equal to 1 for the asset stream ($0 \leq \alpha_i \leq 1$, for each i), by assuming the actual prices being equal to the equilibrium prices and by constraining the intermediation volumes at a level equal to the present value of the liabilities, that is $C_a = C_b = P(t; \boldsymbol{y})$.

---

[27] A specific application to Italian government securities portfolios of this framework with the objective function (31) is proposed by (Mottura 1992) using the one factor Cox, Ingersoll and Ross model. The quadratic problem is analyzed in (Castellani 1991) from both the theoretical and operational point of view.

[28] (De Felice and Moriconi 1991b, pp. 240–263), (Castellani, De Felice, Moriconi and Mottura 1993)

[29] The variable τ, the size of the rebalancing horizon, assumes an important strategic role. Its specification generally depends on management and market factors (operating costs of rebalancing, transaction costs, portfolio's maturities distribution, shocks on leading market variables, administrative deadlines, etc.).

[30] See (De Felice and Moriconi 1990).

In order to cover the fixed liability stream y, an asset portfolio x^* is selected which meets (by assumption) the present value constraint $P(t; y) = P(t; x^*)$, is immunized against additive shifts of the yield curve, and carries minimum risk with respect to an arbitrary interest rate change. The stream $x - x^*$ is the residual asset stream after immunization. It has non-negative cash flows and its present value is equal to the net present value of the outstanding portfolio: it can therefore be interpreted as an "appropriation" stream (consisting of margin, equity capital, and yield on equity capital) which is the surplus stream "disposable" after an optimally immunized nucleus of the outstanding portfolio has been guaranteed. Obviously, with heavily unbalanced intermediation portfolios the surplus stream may not be identifiable: this occurs when the optimization problem has no solutions (infeasible problem)[31].

The appropriation stream could also be defined considering stochastic flows in the asset-liability portfolio, e.g. by using the stochastic problem (31) or (32), provided that the selected stochastic payments are expressed through a properly defined "deterministic equivalent"[32].

Example. Let us consider the outstanding intermediation fixed income portfolio, denominated in "lire", at a given date t.

Specifically, we considered 17 types of asset products (loan) and 10 liability products (bonds). The streams are aggregated by type over a semi-annual time grid and entered into the rows of matrix A (assets) and B (liabilities). Figure 1 shows the profile of the aggregate stream x (asset), y (liability) and of the margin $x - y$.

Given the interest rate structure (assumed to be "expressive of the market"), value and duration gaps may be derived: it results $P_N(t) > 0$, $D(t, x) < D(t, y)$. For the fixed liability stream y, an asset portfolio x^* is chosen which meets (by assumption) the present value constraint $P(t, x^*) = P(t, y)$, is immunized against additive shifts in the yield curve and carries minimum risk with respect to an arbitrary interest rate change.

Figure 2 provides a comparison between streams x and x^*. In Fig. 3 are compared the stream of the outstanding margin $x - y$ and the surplus stream $x - x^*$ implicit in the outstanding assets.

Gap Management Plans. The management of present value and duration gaps that could characterize an outstanding intermediation portfolio can be also approached as linear programming problems similar to (20), or as stochastic asset liability management problem with objective function (30), (31) or (32).

Referring to the previous example, the liability matrix B consists of the aggregate stream y of the outstanding liabilities in the first row, of the stream

---

[31] See (Mottura 1993) for an analysis of surplus stream applied to the management of profit-sharing policies.

[32] The analysis of surplus stream in the case of intermediation portfolios consisting of floating rate bonds and callable bonds (retractable or extendible) are discussed in (Mottura 1992).

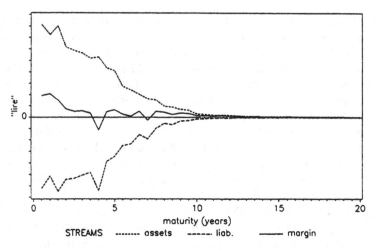

Fig. 1. Payment streams of the outstanding portfolio

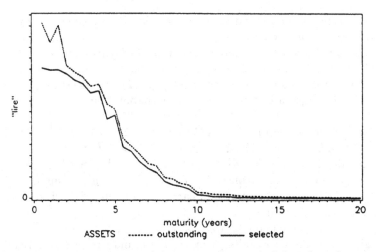

Fig. 2. Payment streams of the outstanding and of the selected asset portfolio

$x - x^*$ (surplus stream) in the second row, and of the streams characteristic of the new fund raising possibilities in the other rows. Similarly, the first row of matrix A contains the aggregate stream x of the outstanding assets and in the other rows the streams which are characteristic of the new investment opportunities.

The asset-liability management problem can be solved after establishing the strategic amount of new assets.

The outstanding situation can be incorporated in the rebalancing problem and the surplus stream can be "protected" by imposing composition constraints equal to 1 for the streams y, x ($\alpha_1 = \beta_1 = 1$) and $x - x^*$ ($\beta_2 = 1$).

By considering problem (20) it is possible to analyze the trade-off, for portfolios immunized against additive shifts, between the volume of new investments and the interest rate risk induced by arbitrary shifts in the curve. It is also

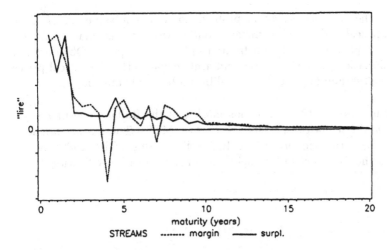

Fig. 3. The outstanding margin and the surplus stream

possible to study the composition of the optimally immunized intermediation portfolios for different liquidity amounts, $C = C_b - C_a$, given the volume of new investments, and hence identify the empirical trade-off between liquidity and standard time spread of the efficient portfolios.

Efficient frontiers of strategic management variables may also be analyzed in the stochastic framework using, for example, problems (31) and (32).

Identification of Market Efficient Frontier. By Theorem 3 we may consistently determine the efficiency locus of the minimum cost immunized portfolios that can be selected from a given matrix A of IRS contracts.

For a given strategic investment period H, the portfolio α, "solution" to the problem

$$\min \sum_{i=1}^{n} \alpha_i q_i^{(a)} \, , \tag{33}$$

$$\text{s.t.} \sum_{i=1}^{n} \alpha_i P(t; a_i) = v(t, t + H) \, , \tag{34}$$

$$\sum_{i=1}^{n} \alpha_i P_r(t; a_i) = v_r(t, t + H) \, , \tag{35}$$

$$\overline{\alpha}_i \leq \alpha_i \leq \overline{\overline{\alpha}}_i \, , \quad i = 1, \dots, n \, , \tag{36}$$

is the efficient portfolio, i.e. the minimum cost immunized portfolio for the period H. It represents, in this sense, a "benchmark" portfolio derived from the matrix A. If A is made of all the contracts traded in the market, α is the efficient market portfolio given the available choices (36), i.e. the minimum cost "synthetic bond" obtainable from the market at that level of duration and taking into account the constraints defining feasible choices (compositions).

By solving the portfolio selection problem for different values of H, it is possible to empirically identify an efficient frontier. Using the normalized prices, it is possible to represent the efficient frontier in the (D, Π) plane. Obviously, in the idealized situation of market prices uniformly equal to the theoretical prices the efficient frontier and the (market) equilibrium line will coincide.

Example. Let us consider the continuous market of the Italian government securities (*Mercato Telematico dei Titoli di Stato* – MTS). The market financial analysis is based on the one factor Cox, Ingersoll and Ross (CIR, 1985) model, which assumes the dynamics for the spot rate (15) with the specification

$$f(r(t), t) = \alpha(\gamma - r(t)) \ , \quad \alpha, \gamma > 0 \ ,$$

$$g(r(t), t) = \rho\sqrt{r(t)} \ , \quad \rho > 0 \ ,$$

and the price of risk

$$q(r(t), t) = \frac{\pi\sqrt{r(t)}}{\rho} \ , \quad \pi \text{ constant.}$$

The valuation equation takes the form:

$$P_t + \alpha\gamma - (\alpha - \pi)rP_r + \frac{1}{2}\rho^2 rP_{rr} - rP = 0 \ . \tag{37}$$

Solving the evaluation equation with the proper boundary condition, it is possible to determine the price of any IRS financial contract and the corresponding risk level. In particular, for $P(t) = v(t, s)$ and under the boundary condition $v(s, s) = 1$, (37) yields (Cox, Ingersoll and Ross, 1985 p.393):

$$v(t, s) = A(s - t) e^{-r(t) B(s-t)} \ ,$$

where the functions A and B are independent of r. This expression provides the equilibrium market line prevailing at time t.

In this particular application we will refer to zero coupon bonds (BOT), coupon bonds (BTP), indexed bonds (CCT), and extendible bonds (CTO). The expression for $v(t, s)$ is sufficient for the derivation of the price and risk of deterministic payment streams. Moreover, the function $v(t, s)$ can be used to estimate the risk-adjusted parameters $\alpha\gamma$, $\alpha - \pi$, and ρ, and the current value of the spot rate $r(t)$ by performing a non-linear regression on the market prices of fixed rate bonds observed on the same date[33]. We performed this estimation on the observed prices of BTPs quoted on November 26, 1991. The results of the estimation procedure are the following:

---

[33] For the details of the estimation of the one factor Cox, Ingersoll and Ross model and the evaluation of IRS securities, with application to the Italian market, see (Castellani, De Felice, Moriconi 1990), (De Felice, Moriconi 1991a).

$$\alpha\gamma = 0.036951 \ ,$$
$$\alpha - \pi = 0.3352 \ ,$$
$$\rho = 0.072522 \ ,$$
$$r(t) = 0.10647 \ .$$

The mean square error of residuals is 0.206 lire, prices refer to a face value of 100 lire and are net of accrued interest. In this case too we impose finite lower bounds to the portfolio weights, α_i, of individual assets. This avoids "exploding" solutions in problem (33). In fact, if underpriced bonds are traded on the market and if $\alpha_i \equiv -\infty$, the investor could create a money pump by selling short unlimited quantities of these bonds. We performed a market efficiency analysis by disallowing short sales, that is by posing $\overline{\alpha}_i = 0$ and $\overline{\overline{\alpha}}_i = +\infty$, $i = 1, 2, \ldots, n$, in the portfolio selection problem. Under these conditions, we have feasible solutions for $\overline{D} \le H \le \overline{\overline{D}}$, where $\overline{D}$ and $\overline{\overline{D}}$ are the minimum and the maximum value, respectively, of the stochastic duration of bonds traded on the market on the valuation date. On November 26, 1991 it was $\overline{D} = 0.18$ and $\overline{\overline{D}} = 4.10$ years.

We have chosen a grid of 32 equally spaced values of H inside this time interval and solved the corresponding linear programming problems. In this application, the theoretical prices of CCTs have been derived adjusting the model for fiscal effects (De Felice, Moriconi and Salvemini 1993). The resulting empirical efficient frontier is illustrated in Fig. 4, where each "square" represents an optimally immunized portfolio (in the stochastic sense).

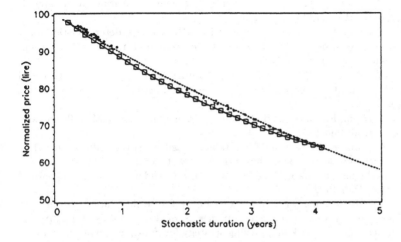

Fig. 4. Market equilibrium line and the efficiency frontier

References

Bierwang, G.O.: Duration analysis. Cambridge, Ballinger Publ. co. 1987

Bierwang, G.O., Kaufman, G.G., Toevs, A.: Recent developments in bond portfolio immunization strategies. In Kaufman, G.G., Bierwang, G.O., Toevs, A. (eds.): Inovations in bond portfolio management: duration analysis and immunization. London, JAI Press 1983 105–157

Bodie, Z.: Pension Funds and Financial Innovation. NBER Working Paper No. 3101 (1989)

Bodie, Z.: Pensions as Retirement Income Insurance. J. of Economic Literature **28** (1990) 28–49

Borch, K.: The optimal portfolio of assets in an insurance company. Trans. of the 18th International Congress of Actuaries (1968) vol. 3, 21–31

Boyle, P.P.: Immunization under stochastic models of the term structure. J. of the Inst. of Actuaries **105** (1978) 177–187

Castellani, G.: Minimum Risk Portfolio Selection in Stochastic Immunization Framework. Unpublished manuscript, 1991

Castellani, G., De Felice, M., Moriconi, F.: Price and Risk of Variable Rate Bonds: An Application of the Cox, Ingersoll, Ross Model to Italian Treasury Credit Certificates. Proceedings of the 1st AFIR International Colloquium (Paris, France 1990) 287–310

Castellani, G., De Felice, M., Moriconi, F.: Asset-liability management. Semi-deterministic and stochastic approach. Trans. of the 24th International Congress of Actuaries (1992a) vol. 2, 35–55

Castellani, G., De Felice, M., Moriconi, F.: Single Factor Diffusion Models for the Global Analysis of the Italian Government Securities Market: Normalized Prices and Efficient Frontier. Research Group on "Models of the Term Structure of Interest Rates", Working Paper n. 6 (1992b)

Castellani, G., De Felice, M., Moriconi, F., Mottura, C.: Un corso sul controllo del rischio di tasso di interesse. Bologna, Il Mulino 1993

Cox, J.C., Ingersoll, J.E., Ross, S.A.: A Theory of the Term Structure of Interest Rates. Econometrica 53 (1985) 385–407

Daykin, C.D., Hey, G.B.: Managing uncertainty in a general insurance company. J. Inst. of Actuaries **117** (1990) 173–277

De Felice, M., Moriconi, F.: Controlling Interest Rate Risk. A semideterministic Model for Asset-Liability Management. Center for Research in Finance - IMI Group, Rome, Working Paper n.4 (1990)

De Felice, M., Moriconi, F.: Uno schema per la valutazione e la gestione di titoli del debito pubblico. In: Ricerche applicate e modelli per la politica economica, Banca d'Italia 1991a

De Felice, M., Moriconi, F.: La teoria dell'immunizzazione finanziaria. Modelli e strategie. Bologna, Il Mulino 1991b

De Felice, M., Moriconi, F., Salvemini, M.T.: Italian treasury credit certificates (CCTs): theory, practice and quirks. BNL Quarterly Review **185** (1993) 127–168

Fabozzi, F.J., Modigliani, F.: Capital markets: institutions and instruments. Englewood Cliffs, Prentice Hall 1992

Fisher, L., Weil, R.L.: Coping with the risk of interest rate fluctuations: returns to bond holders from naive and optimal strategies. J. of Business **44** (1971) 408–431

Fong, H.G., Vasicek, O.A.: A risk minimyzing strategy for multiple liability immunization. Unpublished manuscript 1982

Fong, H.G., Vasicek, O.A.: A risk minimyzing strategy for portolio immunization. J. of Finance **39** (1984) 1541–1546

Granito, M.R.: Bond portfolio immunization. Lexington, Lexington Books 1984

Haynes, A.J., Kirton, R.J.: The financial structure of a life office. Trans. Fac. of Actuaries **21** (1952) 141–218

Ingersoll, J.E.: Theory of financial decision making. Totowa, Rowman & Littlefield 1987

Ingersoll, J.E., Skelton, J., Weil, R.L.: Duration forty years later. J. Fin. Quant. Analysis **13** (1978) 627–650

Klotz, R.G.: Convexity of fixed-income securities. New York, Salomon Brothers 1985

Kocherlakota, R., Rosenbloom, E.S., Shiu, E.S.W.: Algorithms for cash-flow matching. Trans. Soc. of Actuaries **40** (1988) 477–484

Kopprash, R.W.: Understanding duration and volatility. New York, Salomon Brothers 1985

Mari, C.: La teoria dell'immunizzazione finanziaria su reticoli binomiali. Atti del XVI Convegno A.M.A.S.E.S. (Treviso, Italy 1992) 503–517

Matrigali, P., Pacati, C.: Strutture per scadenza dei tassi di interesse come soluzioni duali di problemi di copertura. Atti del XVII Convegno A.M.A.S.E.S. (Ischia, Italy 1993) 625–647

Montrucchio, L., Peccati, L.: A note on Shiu-Fisher-Weil immunization theorem. Insurance: Math. and Ec. **10** (1991) 125–131

Moriconi, F.,: Immunizzazione semideterministica di portafogli di CCT. Gruppo di Ricerca su "Modelli di struttura a termine dei tassi di interesse". Working Paper n. 5 (1991)

Mottura, C.: Uno schema di "asset-liability management" per contratti finanziari con flusso di pagamenti aleatorio. Atti del XVI Convegno A.M.A.S.E.S. (Treviso, Italy 1992) 547–562

Mottura, C.: Managing profit-sharing policies in a financial immunization framework. Proceedings of the 3rd AFIR International Colloquium (Roma, Italy 1993) 807–819

Redington, F.M.: Review of the principles of life office valuations (with discussion). J. Inst. of Actuaries **78** (1952) 286–340

Shiu, E.S.W.: A generalization of Redington's theory of immunization. Actuarial Research Clearing House **2** (1986) 69–81

Shiu, E.S.W.: Immunization. The matching of assets and liabilities. In MecNeill, B., Umphrey, G.J. (eds.): Actuarial Science. Dordrecht, Reidel Publ. co. 1987 145–156

Financial Risk, Financial Intermediaries and Game Theory.

Flavio Pressacco

Dipartimento di finanza dell'impresa e dei mercati finanziari. Università,
Via Tomadini 30, 33100 Udine, Italy

1 Introduction

Financial intermediaries exist and were already studied well before the second half of our twentieth century. Yet in the last fifty years many new and more precise insights on financial institutions have been derived exploiting results obtained by researchers of the scientific area defined (more or less precisely) as modern-quantitative finance.

It is a widely shared opinion that the modern quantitative theory of finance is based on two strong theoretical pillars.

The first pillar is that of portfolio selection and risk return based asset pricing models:it was built in the fifties and sixties with the overhelming contributions of the Nobel Prize winners H. Markovitz (Markovitz 1952), and W. Sharpe (Sharpe 1964). Two fundamental financial laws were the result of that wave:

a) no (excess) profit without risk,
b) rewards only for the efficient risk (and not for any risk).

According to this approach the main role of financial institutions was to provide efficient portfolios for the saving units, and competition among intermediaries was based on the effort to be more than efficient, or more precisely to obtain better rewards than those normal for the relevant riskiness of the intermediary's portfolio.

Thousands of papers aiming at defining efficiency properly and at obtaining empirical evidence about the ability of intermediaries to be more than efficient (or to beat the market) have indeed appeared since then.

The second pillar, that of arbitrage free pricing theory, came in the seventies with the celebrated papers by Black-Scholes (Black-Scholes 1973), Merton (Merton 1973) and Ross (Ross 1976), followed by the later refinements of Harrison-Kreps (Harrison-Kreps 1979) and Harrison-Pliska (Harrison-Pliska 1981) and is currently proceeding with an astonishing number of new applications.

In its very essence the arbitrage free pricing theory offered instruments to realize, under rigorously defined conditions, the old dream of obtaining profit without risk.

Accordingly, the role of intermediaries became that of completing the market and exploiting the ability to create synthetic assets, netting a sure trading profit in the case of mispricing between the synthetic assets and the original ones. There is no need to underline here that organized markets for such financial assets as options, futures, swaps and so on, received a big impulse from the arbitrage free approach.

Yet, despite these successful insights, we are far from being able to give a complete and formalized satisfactory explanation of the role of financial intermediaries in financial markets and more generally in economics: probably a third pillar is needed to this purpose, and in my opinion this is precisely the pillar of game theory applications in finance.

The basic idea behind this third pillar is that another essential role of a financial intermediary is that of (delegated) screening and monitoring of entrepeneurial projects on behalf of saving units in a world where asymmetric information and moral hazard are the rule rather than the exception.

More precisely the screening role tries to solve the problem of ex ante asymmetric information, whereas the monitoring role aims to solve that of ex post asymmetric information about the outcome of entrepeneurial projects. In principle an intermediary exploiting scale economies is able to do these jobs better than the community of saving units, but the point becomes now: who does screen and monitor the intermediary? (in our Latin language here in Rome: quis custodies ipsos custodes?). And under what conditions might there be an incentive to fair non opportunistic behavior on the part of intermediaries, thus giving rise to a reliable self certification of screening and monitoring?

In my opinion game theoretic models seem to be able to provide an adequate formalized setting of obtaining deeper insights into these interesting problems.[1] This is the core of the following reflexions.

The paper is organized as follows: the first section (Chapts. 2–5) is devoted to a short introduction on the role of financial intermediaries, that essentially tries to summarize in a convenient formal language results well known in the economic literature. A second section (Chapts. 6, 7) is devoted to a brief recall of game theory terminology useful in the applications: the relevant area is the one of two person non cooperative games of incomplete information in extensive form, with attention paid to the sophisticated refinements of the by now classical Nash equilibrium which has recently appeared in the specialized literature. In the final section (Chapts. 8–12) some examples of "finance games" concerning the role and behavior of financial intermediaries are introduced and briefly discussed.[2]

---

[1] There is in nowadays a widespread agreement that the economics of asymmetric information and moral hazard (even not necessarily within a strict game theoretic approach) is the winning approach to a modern theory of financial intermediaries. As regards the domestic debate in Italy see for example, Ciocca et. al. (1991), Forestieri (1993), Fulghieri e Rovelli (1993), Masera (1991).

[2] Without any claim at being exaustive or updated in the choice of the examples: in fact some of the papers quoted in the references could have deserved adequate attention, too.

Section 2

Schematically (see Fig. 1), a financial intermediary is an institution that receives money from savers against a promise of monetary performances whose total sum is normally contingent to the occurrence of given states of nature.

In return for such a promise, savers pay a price P. This sum is generally invested in a portfolio of assets aimed at generating the resources required to honour the liabilities the intermediary has taken on.Often the intermediary also possesses his own funds, G which, when opportunely invested in a similar fashion, generate additional resources that might prove necessary to cover the liabilities, were the income from the investment of the price paid by the savers to turn out insufficient to this end.

In the unfortunate case that at the maturity of the liabilities the intermediary does not have sufficient resources to honour them, he will partially or totally be insolvent.

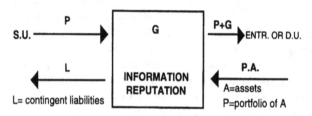

Fig. 1.

Section 3

Continuing with our schematic presentation, the committments of the intermediary may be defined in three conceptually different ways:

a) fixed (in monetary terms) face value committments, independently of the state of nature that occurs;

b) contingent face value committments connected to the effect of variables beyond the control of the intermediary, that is contingent to the occurrence of observable states of the world that cannot be influenced by the decisions of the intermediary;

c) contingent face value commitments but which depend completely or in part on the actions of the intermediary, that is contingent to states of the world that may be influenced by the intermediary (for example the performance of a portfolio freely managed by the intermediary without precise limits as to its composition).

The three above definitions entail financial risk of a different nature, both for the savers and for the intermediary. To gain some insight into this point we shall take as a starting point an elementary text-book case of a very simple economy.

To be precise the economy is a single period two states (binomial), that is an economy in which at the end of the period only two states of nature are possible: w_1 and w_2; two original assets are negotiated in this economy:one risk free with a sure yield r and one that is risky with a random yield X, whose value is u in the first state of the world and d in the second. Without losing generality, let us suppose that $u > r > 1 > d$. The economy is governed by arbitrage free pricing rules.

Section 4

Let us briefly recall the arbitrage free pricing rules that operate:

a) there exists a risk adjusted probability π of state w_1 (and therefore a corresponding $1 - \pi$ of state w_2), such that each contingent claim (c_1, c_2) payable at epoch 1 has an arbitrage free price at epoch 0 equal to

$$P(c_1, c_2) = (1/r)(\pi c_1 + (1 - \pi)c_2) \tag{4.1}$$

that is, it may be expressed as the present value at the risk free rate of the expectation, with respect to the risk adjusted probability, of the contingent claim.

b) the risk adjusted probability is derived from the condition

$$\pi u + (1 - \pi)d = r \tag{4.2}$$

that it is equal to

$$\pi = (r - d)/(u - d) \tag{4.3}$$

In a meaningful geometrical interpretation this signifies that the prices of all the contingent claims must be such that the points corresponding to their contingent yields lie on a straight line passing through the points of coordinates (r, r) and (u, d). The slope of this line is: $-\pi/(1 - \pi)$. This line could be called the characteristic line of the economy. (see Fig. 2).

With reference to this economy let us compute the arbitrage free price of the liabilities of a financial intermediary, liabilities that could be seen as a special type of contingent claim.

The liabilities of the intermediary derive from a contract stipulated at the beginning of the period (epoch 0), establishing a payment of a price P on the part of the saver, against the promise of the payment of a sum L (contingent to the state of nature) at the end of the period (epoch 1), effected by an intermediary initially possessing his own funds for an amount G.

Let us treat the case where the face value of the liability is given, (with m a constant, agreed between the intermediary and the saving unit) by

$$L(X) = mX + (1 - m)r \tag{4.4}$$

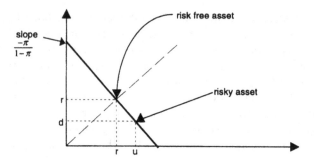

Fig. 2.

which amounts to saying that we have liabilities of the second type (contingent to an index exogeneous to the intermediary's control), with the proviso that the corresponding payment is however contingent to the solvency of the intermediary, so that it cannot be in any case greater than the resources of the intermediary at epoch 1, which in turn are generally a function of three factors: the price P paid by the saver, the intermediary's initial own resources G, and the intermediary's investment policy, summarized here by the value α percentage of resources invested in the risky asset.[3]

Formally this amounts to saying that the effective value of the intermediary's liability, keeping account of insolvency, is given by

$$\min(mX + (1-m)r, (G+P)(\alpha X + (1-\alpha)r)) \tag{4.5}$$

It may be checked with no difficulty that there ensues total lack of risk of insolvency (and hence $P = 1$ for each m lesser than α!) for values of the guarantee which are not below the critical level

$$G* = ((\alpha - m)(r - d))/(\alpha d + (1-\alpha)r) \tag{4.6}$$

Some tedious calculations give us the following expression for the arbitrage free price as a function of G when $G < G*$

$$P(G) = (r^{-1}(\pi(mu+(1-m)r)+(1-\pi)G(\alpha d+(1-\alpha)r)))/(1-r^{-1}(1-\pi)(\alpha d+(1-\alpha)r)) \tag{4.7}$$

Then the arbitrage free pricing function $P(G)$ turns out to be piece-wise linear, continuous, rising up to $G*$ and then constant with P definitively equal to 1. Note that when $m = 0$ the results refer to a fixed non contingent face value r of the liability (that is to liabilities of the first type).

Summing up, the role of the guarantee fund G is that of allowing the intermediary to adopt investment strategies of P (and G) riskier than those on which the liability contract is based. When the guarantee fund reaches the critical level $G*$ the saver is fully guaranteed. Lower levels of G involve a risk of partial insolvency in the bad state (a well known result in the actuarial world, even without the

---

[3] It is further supposed that α is greater than m so that the investment policy of the intermediary is more aggressive than that implied by the liability rule $L(X)$.

arbitrage free hypothesis!), the difference between the face value of the liability and the amount actually available increasing with the difference $G*-G$.

Clearly the greater the amount of the potential insolvency the lower is the price to be paid, or said another way, with some probability of defaulting the price is lesser than the (full) price of the same face value liability without default risk. (Note that the price to be paid is always "fair" in as much as it is arbitrage free).

Roughly summarizing, the alternatives open to the intermediary are the following:

a) full prices with high risk of the investment,(as captured by the *alpha* value, or by the difference $\alpha - m$) offset by high values of the guarantee fund;
b) full prices, average investment risk and average guarantee fund, which is nevertheless sufficient to avoid the saver running any risk;
c) an investment risk which is higher than that covered by the guarantee fund in exchange for a reduction in prices, with a partial shift of the risk onto the shoulders of savers.

Finally let us deal with contracts of the third type where liabilities are linked in some way (also) to the performance of the intermediary.

The "pure" case in which the face value of the liability is:

$$P(\alpha X + (1 - \alpha)r) \tag{4.8}$$

with the intermediary's investment parameter set at α is banal: the arbitrage free price is obviously P.

Less obvious cases would be mixtures of contracts of the second and third type; it is not necessary to treat the question here.

Section 5

Now let us go back to analyze contracts between a saving unit and an intermediary. As for contracts of the first two types it is evident that the financial risk of the saver could be split into three parts: the first comes from the desired risk (the most preferred position) as revealed by the choice of the parameter m in the L function 4.1.

The second comes from deviations in the investment strategy of the intermediary (α) from the preferred one (m) provided that these deviations are not fully guaranteed by a high enough value of G, so that the face value of the claim is different from the real value (bearing in mind the solvency risk). This could be succintly defined as fair default risk because the price is supposed to be the arbitrage free price.

Finally a third source of risk comes from unfair (opportunistic) behavior on the part of the intermediary that implements riskier strategies (or alternatively lowers the guarantee fund) than those implicitly or explicitly used to charge the arbitrage free price.

For the intermediary this additional risk is a source of (unfair) extra profits.

It turns out that an opportunistic behavior is to be expected when the values of G and α are not observable, or, even in case of observability, if there is no enforceability of the contracts (i.e. proper punishment mechanisms in the case of deviations from the contractual conditions).

To summarize the viability of contracts of the first set requires observability of the couple (G, α) and enforceability of the related contracts, while the viability of contracts of the third type depends on the observability of returns of the asset portfolio of the intermediary.[4]

Looking now at the conditions prevailing in the real world, with great difficulties even for specialized intermediaries and a fortiori for single saving units in monitoring the exact values of the relevant variables, strong incentives toward an opportunistic behavior of intermediaries seem to prevail.

The likely reaction of the saving units should be to implement costly mechanisms of monitoring,[5] yet recent events in financial markets have shown that

---

[4] Fundamental papers treating the role of financial intermediaries, with deep reflexions on the viability of financial contracts are those of Diamond (Diamond 1984) and Leland-Pyle (1977).

[5] In more detail, bearing in mind that there are strong incentives for the intermediary to be opportunistic and gain unfair extra profits, savers should react in the following ways:

1) choose contracts of the third type supported by (costly) mechanisms granting precise ex post observability of the performances of the intermediary;

2) require a (n ex ante observable as a precondition of contractual validity) guarantee level G, that is so high that, even with the riskiest possible strategy chosen by the intermediary, solvency is fully granted (this level would be much higher than the one needed according to the agreed contractual strategy).

Both solutions have some counter arguments as 1) implies, quite probably deviations from the preferred portfolio position of the saver, and, as remarked, costly mechanisms of observability, and 2) demands solvency margins well above that required by economic efficiency, or costly control mechanisms enforced by some supervisory authority.

With particular reference to contracts of the third type, it should be underlined that observability of returns depends in turn on the quality of assets that enter the fund (the best quality is day to day quotation in some official exchange or in telematic markets) jointly with procedures that render it possible to check the composition of the fund.

As for the inability of the saver to control the risk quality of the dedicated portfolio managed by the intermediary, two remarks are in order: some commitment of the intermediary to follow (at least with some degree of freedom) an announced style (see Sharpe 1991) may help to identify the risk quality of the fund, exploiting at the same time the ability of active managers of the fund to make use of superior ability of timing and/or selection to beat the market or at least to beat a passively managed fund. A tradeoff between some likely deviation from the best preferred risky position of the saver, justified by the ability of the intermediary to implement an active (and cheap!) strategy to beat the market, makes contracts of the third type viable and convenient.

even costly and apparently sophisticated monitoring procedures and structures are far from granting complete or at least acceptable reliability.

A dramatic picture of a failure of some financial markets, looking dangerously like lemon markets (Ackerlof 1970), seems to emerge from this line of reasoning, so that, to quote another Nobel Prize winner: "we should be impressed that financial markets work as well as they do, despite natural impediments" (Greenwald and Stiglitz (1989) pag. 6).

Whatever the case, it is apparent that saving units need (explicit or implicit) help in their effort to monitor financial intermediaries. Schematically five tentative answers have been selected in the evolution of the financial world, and of course nothing prevents it from using a proper mixture of all mechanisms:

a) monitoring by higher level financial institutions, acting, with some power, as regulatory authorities;

b) monitoring by independent agencies;

c) peer monitoring;

d) mechanisms of a priori self certification (signalling) by financial institutions;

e) incentives to self control on the part of financial intermediaries, based on the concept of marketable reputation.

The last (but not least) two points can be studied and formalized in a proper game theoretical setting, but the ideas behind the formalized approach are very simple and go back to the pioneering paper of Leland-Pyle (Leland-Pyle 1977). The authors suggest that to help screening between borrowers, with private information about the expected profitability of their asset portfolio, a **signalling** device should be involved: borrowers with better perspectives should credibly signal their optimistic beliefs, and it is suggested that the best signal to this purpose is to invest a part of the borrower's wealth in the project.

Two comments are in order: first a complementary role of the guarantee fund G emerges beyond that of granting payments of the liabilities in the event of poor performances of the asset portfolio, indeed the guarantee is used to increase the (lender's) beliefs (probability) of good performances of the borrower.

Second when speaking of wealth we should intend not merely money or tangible assets but intangible assets too.

And indeed the opportunity to play with intangible assets like **reputation**, is a crucial point in the birth of new intermediaries, or intermediaries entering a new field.

Yet, as we can observe daily, financial intermediaries exist (commercial banks, merchant banks, life insurance companies, investment funds, pension funds....) which after all do not play just a one stage game.

On the contrary it is precisely the survival and the hope of continuing to be active with some profitability in the future that accounts for fair behavior on the part of the intermediaries, thus avoiding the failure of the market.

Section 6

As said in the introduction the relevant area is here the one of two person non cooperative games of **incomplete information** in **extensive form**.[6]

The two players are a financial institution and a saving unit; the financial institution "borrows" (in a broad sense, that is with contracts of one of the three types discussed in para 3) money from the saving unit, selects an asset portfolio and then, if possible, repays the lender.

Moreover there is incomplete information in the sense that the financial institution is of a **type** not known by the saving unit, and different types may adopt different strategies (i.e. select different portfolios) during the game, because they have different preferences (payoffs).

In game theory terminology a game in extensive form is described by the so called game tree. The **game tree** (or the extensive form representation) specifies:

a) when each player has the move,
b) what each player can do when he has the move,
c) what each player knows when he has the move,
d) the payoff received by each player for each combination of moves chosen by the players.

In games of incomplete information usually the game tree begins at a starting decision node, where a third player **"nature"** has the introductory move, that consists in the choice, with some probability that is public information, of a type t of the player of unknown type (the financial institution) among a set T of possible types. This choice is immediately revealed only to the financial institution and not to the other player (the saving unit).

A strategy for a player is a complete plan of action, that is a plan that specifies a feasible action (that may well be a random action) in every contingency in which the player (the type) might be called on to act.

An information set for a player is a collection of decision nodes such that:

(i) the player has the move at every node in the set and
(ii) when the play of the game reaches a node in the information set, the player with the move does not know which precise node has been reached. It is standard to denote that a collection of decision nodes are an information set by connecting the nodes with a dotted line.

After that a strategy is a plan that specifies a feasible action for any information set of a player.

It is well known that a key concept in game theory is the **Nash equilibrium**; succintly in a two person game a Nash equilibrium is a couple of strategies (one for each player) that are reciprocal best replies (or at least that belong to the set of reciprocal best replies in case the set of best replies of a player is not a singleton).

---

[6] In this section we will largely have recourse to the terminology suggested in Gibbons' book (Gibbons 1992).

Section 7

To be tailored to the sector of extensive form games of incomplete information this definition needs some refinements. It is intuitive that the best choice of a player at a given information set depends on his beliefs about which node of the information set has been reached. And indeed a perfect **Bayesian-Nash equilibrium** for our extensive form games of incomplete information is a system of **strategies** of players and **beliefs** (at any information set and not merely at information sets that are expected to be reached through equilibrium playing), such that the players' strategies are **sequentially rational**, and **beliefs and strategies** are **coherent** in a convenient sense.

Precisely, a system of strategies and beliefs is:

(i) sequentially rational if at any information set the action taken by the player with the move is optimal given his beliefs at that information set and the subsequent strategies of that player and of the other players (that define unequivocally, perhaps randomly, the behavior of the players in any future contingency);

(ii) coherent if at information sets on the equilibrium path (that is a set that could be reached with some positive probability if the players use their equilibrium strategies), beliefs are determined by Bayes' rule and the players' equilibrium strategies, while at information sets off the equilibrium path (that is with zero probability of being reached under equilibrium playing) beliefs are determined, whenever possible, by Bayes' rule and the players' equilibrium strategies, and are unconstrained otherwise.

For a better understanding of the given definitions, let us consider a very simple example of a **signalling game**, that is of a game belonging to an important class of two person extensive form games of incomplete information.

The game tree is shown in Fig. 3.

The game starts at the white node with nature's choice of the type of the first player (let us say a bank) generally named sender. Nature chooses type t_1 or B_1 with probability ρ that is public information and of course type $t_2(B_2)$ with probability $1 - \rho$.

The bank immediately observes its type, and then recognizes that it finds itself in a singleton information set (at B_1 or B_2) and selects an action (named a message, because, as we shall see, it may convey some information about its type) from a set of (two for both types in our example) feasible messages S_1 or S_2.

The other player (let us say a saver) named the receiver, observes the message, but not the type; so he recognizes he is in information set I_1 (or respectively I_2),but does not know which node of the information set has been reached. Nevertheless he must select an action from a set of feasible actions (left or right in the example for both information sets).

The expected payoff for both players are functions $g_1(t, s, a)$ and $g_2(t, s, a)$ of the moves of the three players: nature, sender and receiver.

GAME TREE

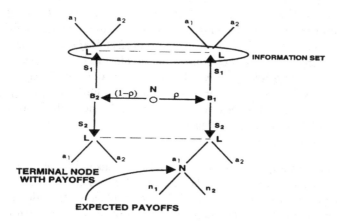

Fig. 3.

Next let us concentrate on the beliefs of the receiver at his two possible information sets; suppose for the moment that an equilibrium strategy of the bank is to play s_1 with probability p at information set B_1, and to play s_1 with probability q at the information set B_2 (Fig. 4)

Both the information sets of the saver lie on the equilibrium path unless p and q are both zero or respectively both one.

Then the coherent beliefs are derived by an elementary application of the Bayes' rule: (see the side values corresponding to nodes of the information sets in Figs. 4 and 5).

What about the case $p = q = 1$? The information set I_2 now has zero probability of being reached (is off the equilibrium path) and Bayes' rule cannot help us: beliefs at the information set cannot be coherently derived or at least constrained.(Fig. 6).

Yet note that, even in this case, unrestricted beliefs could rule out a strongly dominated action off the equilibrium path (that is a decision with a lower expected payoff than another feasible action for any probability distribution of the saver at information set I_2).

If they are (part of) an equilibrium, the couple of strategies ($p = q = 1$) are an example of a **pooling equilibrium** (in pure strategies), that is of an equilibrium in which both types send the same message. Conversely if the equilibrium values satisfy $pq = 0$ and $p + q = 1$, the **equilibrium is separating** because each type sends a different message.

The following results clearly hold:

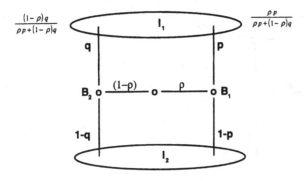

Fig. 4.

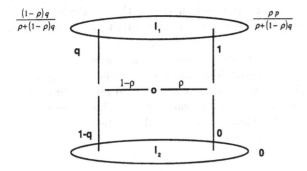

Fig. 5.

(i) in a pooling equilibrium, at the information set on the equilibrium path the receiver's beliefs on the sender's type are not changed with respect to the original ones given by ρ; indeed the message received conveys no information on the sender's type as both types send the same message; look back at picture 6;

(ii) in a separating equilibrium with two messages only (and two types) both information sets of the receiver are on an equilibrium path, and the beliefs of the receiver at an information set give probability one to the type that in equilibrium sends that message, and zero to the other type;

(iii) in a separating equilibrium with more than two feasible messages (and two types), there are information sets off the equilibrium path.

Let us recall here that convenient constraints on beliefs at information sets off the equilibrium path define more sophisticated refinements of equilibrium beyond the

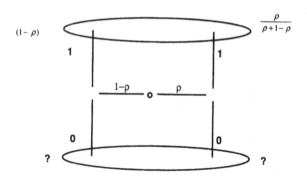

Fig. 6.

perfect Bayes-Nash previously defined. Reasonable restrictions can be derived on the basis of careful reasoning about the advantage of the various type of senders to deviate from the equilibrium (select signals off the equilibrium path). For signalling games an important refinement is the **intuitive equilibrium** (Cho-Kreps 1987).

In signalling games a (pooling or separating) equilibrium is intuitive if at information sets off the equilibrium path it is supported by beliefs that give zero probability to types for which that message is strongly **equilibrium dominated** (that is for any action of the receiver, the message gives to that type a payoff lower than the equilibrium payoff). Clearly this is possible if there is at least one type that is not equilibrium dominated.

Section 8

Let us now pass on to treat a few models of financial games of incomplete information in extensive form, (connected to papers) which have recently appeared in the literature.

The first one is a modified version of a signalling game (Lucas and McDonald 1992),that captures the essential ideas of the original paper, while saving a lot of analytical and sometimes irrelevant trouble.

Let us consider precisely two types of banks, that differ for the value of the guarantee fund G which is private information of the bank. Both types of banks may act as investors on a single period arbitrage free binomial economy such as the one described in Chapt. 4. They may send a signal locking in a given (enforceable) portfolio strategy, putting a percentage α in the risky asset (with yield u in the good state and d in the bad state) and $1 - \alpha$ in the risk free asset.

Next the bank stipulates with a saver a one period deposit contract for a fixed amount P, at an agreed face value yield g, promising payment of one unit at the end of the period. Note that the deposit may be risky, at least for the low quality bank, the one with a low value of G. In other words there is some probability of partial default in the bad state for the low quality bank.

The yield on deposit depends on the beliefs of the lenders about the quality of the bank and is intuitively higher for the low quality bank. Thus it may be that the banks may send a signal of their quality through a convenient choice of their portfolio policy.

Let us look for some meaningful equilibrium of this signalling game by starting to treat the case of a bank whose quality G is public information and that locks in a portfolio strategy α (percentage invested in the risky asset).

Let $f(G, \alpha) = 0$ be the function implicitly defined by Eq.

$$(r^{-1} + G)(\alpha d + (1 - \alpha)r) - 1 = 0 \qquad (8.1)$$

This function defines couples (G, α) such that, with an initial price $P = r^{-1}$, the final value of the bank's portfolio is in the bad state just equal to one, that is just enough to grant solvency in any state.

For values of α between zero and one, explicit functions $G(\alpha)$ or $\alpha(G)$ may immediately be derived, namely:

$$G(\alpha) = \alpha(1 - r^{-1}d)/(r - \alpha(r - d)) \qquad (8.2)$$

$$\alpha(G) = Gr/(G(r - d) + (1 - r^{-1}d)) \qquad (8.3)$$

Both functions are increasing, G is concave ($G''(\alpha)$ is positive), while $\alpha(G)$ is convex.

With reference to a coordinate plane (α, G), the area above the graph of $G(\alpha)$ represents couples of type and strategy such that with an initial price $P = r^{-1}$, the solvency is a fortiori granted in any case so that the arbitrage free price of a loan with unit face value is r^{-1}. Let us call it the solvency area. Conversely if we use the coordinate plane (G, α), the solvency area lies below the graph of $\alpha(G)$. (See Figs. 7, 8).

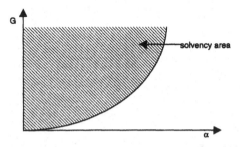

Fig. 7.

What happens outside the solvency area? It could be shown that the arbitrage free price of a single period loan contract with face value of one monetary unit, is given by:

$$P(G, \alpha) = (r^{-1}\pi + (1 - \pi)G(1 - \alpha(1 - r^{-1}d)))/(\pi + \alpha(1 - \pi)(1 - r^{-1}d)) \qquad (8.4)$$

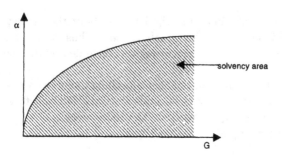

Fig. 8.

The above results help us in defining the behavior of the arbitrage free price as a function $P_G(\alpha)$ of α, for a fixed value of G.

Precisely:

(i) if G is not less than the critical value $G(1) = (r - d)/rd$, $P_G(\alpha)$ equals r^{-1} constant for any α feasible in the interval $(0, 1)$.

(ii) if $G < G(1)$, $P_G(\alpha)$ equals r^{-1} only in the interval $(0, \alpha(G))$,with $\alpha(G)$ defined by 8.2, and is given by 8.4 in the interval $(\alpha(G), 1)$, thus being strictly decreasing in this interval.

As a simple example it is worth studying the case $G' = G(1)$ (high quality) and G'' (low quality) of the possible bank types, so that $P'_G(\alpha)$ is constant at r^{-1} while $P''_G(\alpha)$ equals r^{-1} for α less than $\alpha(G'')$ and is given by 8.4 for bigger values of α. (see Fig. 9)

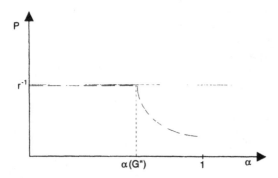

Fig. 9.

Now tedious but straightforward calculations show that in a risk averse market with actual probability p of the good state greater than the risk adjusted one π, the expected profit of the good bank. Precisely the expected profit is defined (with E the expectation operator) as

$$E((r^{-1} + (rd)^{-1}(r-d))(\alpha X + (1-\alpha)r) - 1 - d^{-1}(r-d))$$

turns out to be equal to $d^{-1}(u-d)\alpha(p-\pi)$, that is an increasing function of α.

In turn the rate of return of the same bank is equal to $r + \alpha r(r-d)^{-1}(u-r)$ in the good state and $r - \alpha r$ in the bad state, so that the riskness too increases with α. The best choice α^* of α for the correctly identified good bank depends then on the best risk-return combination:let us suppose that it lies inside the interval $(0,1)$.

Turning now to a consideration of a correctly identified bad bank, a little algebra shows that the final cash flow is zero in the bad state and $Gr(u - d)(r-d)^{-1}$, in the good one, both cash flows being independent from α. This means that the bad bank does not care about the portfolio choice: all signals are equivalent.

Of course this is no longer true if in sending some message α, the bad bank could induce positive beliefs that it is a good bank. Indeed for any α greater than $\alpha(G'')$, being perceived as a good bank gives a net initial positive inflow $(r^{-1} - P(\alpha))$ in excess of the arbitrage free price.

The expected gain accruing to the bank, if the beliefs of being a good bank are denoted by $\rho(\alpha)$ is then given by the expectation of

$$\rho(\alpha)(r^{-1} - P(\alpha))(\alpha X + (1-\alpha)r) \qquad (8.5)$$

In this game a separating equilibrium exists provided that the value of the optimal choice α^* of the good bank (correctly perceived) is smaller than $\alpha(G'')$.

To be precise the equilibrium is $(\alpha^*, \alpha(G''))$, supported by Bayesian beliefs at signals on the equilibrium path, and by beliefs off the equilibrium path that give probability one to the event that any signal off the equilibrium comes from a bad bank.

Indeed we immediately realize that both banks have neither an incentive to leave their own signal and pool the other one, nor to play a signal off the equilibrium.

A more interesting and meaningful model could be obtained by introducing deposit insurance, with say a governmental agency paying a sum, not greater than an upper bound c, in case of default of the bad bank, with a corresponding increase of the arbitrage free price charged by the bank to depositors.

It could be shown that in this case the bad bank obtains an advantage, that pushes the optimal value of α to its maximum value 1(it could be said that for the bad bank it is "costly" to hold riskless securities).

A candidate separating equilibrium is then $(\alpha^*, 1)$, but this fails if the bad bank prefers pooling the good one, sending α^* in order to be treated as a good bank rather than playing 1 and be correctly perceived. In this case a separating equilibrium like the one proposed by Spence in his seminal paper (Spence 1973) may emerge lowering the message of the good bank to that level (if any) that exactly deters the bad bank from pooling. For some convenient values of the parameters this equilibrium is supported by intuitive beliefs at signals off the equilibrium.

Section 9

In the single period model of Lucas-Mc Donald the reputation (of being a good bank) is immediately acquired on giving a signal; there are multiperiod models where on the contrary reputation is built up or enforced as a result of a sequence of proper actions.

The distinctive feature of these models is that the intermediary plays low risky moves (safe securities or low risky loans) that, from a myopic point of view, are worse than other riskier opportunities (high risky loans).

The intuitive idea is that, by doing so, the intermediary trades off between an expected short period of unfair excess profit, but with a high probability of being detected and fired off the market, and a long lasting period of fair profits.

It is not easy to move from intuition to formal models in a proper game theoretic approach. We will comment here on two of these models, both relying on the perfect Bayes-Nash equilibrium concept, yet very different in their logic and in their conclusions.

The first one is the so called credit game (Terlizzese 1988) and attempts to mimick in a financial setting the solution (based indeed on perfect Bayes-Nash equilibrium, or more precisely on **sequential equilibrium**, that is an equilibrium defined in a slightly different way from the perfect Bayes-Nash one) given to the so called centipede game paradox (Kreps-Wilson 1982), a game of incomplete information with a large but finite number of stages.

The multistage credit game is built as follows (see Fig. 10):

(i) nature draws with some probability ρ (which is public information) one of two types of banks;

(ii) there is a sequence of savers, each one playing a single stage of the game. Each saver ignoring the outcome of nature's choice, selects one of two actions: give money to the bank under a risky debt contract with face gross return (yield) r constant at any stage(!) (deposit-play, D), or keep the money (not play, N) with gross return 1.

(iii) the bank knowing the type it is, has the move only if the saver deposits and may choose between a (single period) risky loan (play R), or a (single period) sure asset (play S); both the choice of the bank and the outcome is private information and only in the case of insolvency are they indirectly induced.

(iiii) for the bank the outcome of the strategy S is obviously known in advance, while the outcome of the risky loan is random (another move of nature) with say two possible results: high (H) with some known probability p_H and low (L) with probability $1 - p_H = p_L$.

In case of L, the bank is (perhaps partially) insolvent (defaults) and this is public information for the next generations of savers; in the case of H, the bank is able to pay the face value of its liability but there is still uncertainty about its type.

(iiiii) It is now time to specify that ,with reference to any single stage, the bank of the first type (let us call it a good bank) prefers to play safe (perhaps

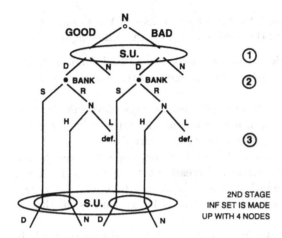

① RISKY DEBT CONTRACT WITH FACE
VALUE r (1ST TYPE)
② IN EACH PERIOD GOOD BANK PREFERS S
THE BAD BANK PREFERS R
③ S.U. OBSERVES ONLY DEFAULT

Fig. 10.

for ethic motivations or some other exogenous reasons), while type two bank
(bad bank) prefers to play risky loans.

It is convenient to begin treating formally a single stage problem. Suppose
the saver beliefs that the bank is bad with probability ρ; anticipating that it will
play risky, while a good bank will play sure, his expected payoff of the deposit
strategy is, with d as the yield received in case of partial default:

$$\rho(p_H r + p_L d) + (1 - \rho)r \tag{9.1}$$

Leaving aside risk aversion problems, suppose that this expectation is greater
than 1, so that he prefers to deposit.

In this case, the single stage game has a perfect Bayes-Nash equilibrium in
pure strategies, supported by beliefs such that the expectation 9.1 is greater
than 1: deposit for the saver, play sure for the good bank and play risky for the
bad bank.

Passing on to an N stage game (with N large but finite), we must stress the
fact that the saver who has the move, has the opportunity to update his beliefs
on the quality of the bank, on the basis of what has been publicly observed in
the previous stages (repayments or defaults). This in turn renders the behavior
of the bad bank no longer straightforward as in the single stage game.

It could be shown that, at least for convenient values of the parameters, a perfect Bayes-Nash equilibrium of the game exists with the following features:

1) the game period may be divided into two parts, say a "body" and a "tail", the length T of the tail being determined by some of the parameters of the game.

2) Except for the good bank, which in equilibrium always plays safe, both in the body and in the tail period, if it is given the move, the behavior of the bad bank and of the savers is different in the two periods.

3) To be exact there are a constant β (lesser than one) dependent on the parameters of the game and a function z_n ,decreasing for n up to $N-T-1$ and then identically equal to 1, which define a "boundary function" $b_n = \beta z_n$, (thus decreasing in the body and constant at β in the tail), such that in equilibrium:

3a) the bad bank always plays risky in the tail if it is given the move, while in the body it either plays sure or randomizes; to be precise, at time n in the body: if $\rho_n > \beta z_{n+1}$

it plays risky with probability

$\phi_n = (\rho_n - \beta z_{n+1})/((1 - \beta z_{n+1})p_L \rho_n)$, and safe with probability $1 - \phi_n$.

while if $\rho_n < \beta z_{n+1}$ it plays safe.

In the above formulae ρ_n is the saver's belief at stage n that he is playing with a bad bank.

3b) the saver's strategy at stage n is:

if $\rho_n > \beta z_n$ he keeps the money stored (does not play).

if $\rho_n < \beta z_n$ he deposits

if $\rho_n = \beta z_n$ he randomizes conveniently.

Apparently there is no difference between the behavior in the body or in the tail, but remember that $z_n = 1$ identically in the tail.

4a) Finally the evolution of beliefs in the absence of previous defaults and provided the saver has played D, is guided by the following Bayesian rule, where ϕ_n is the probability that the bad bank plays risky at stage n:

$$\rho_{n+1} = (\rho_n(\phi_n p_H + (1 - \phi_n)))/((\rho_n(\phi_n p_H + (1 - \phi_n)) + (1 - \rho_n)) \quad (9.2)$$

4b) If the saver did not play in the last stage, there is no new information and coherently $\rho_{n+1} = \rho_n$, while if some default has previously happened, then $\rho_{n+1} = 1$. Note that an information set with a default may be on the equilibrium path (following a randomization by the bad bank in the body, or a risky play in the tail, and in this case the belief $\rho_{n+1} = 1$ comes from the Bayes' rule), as well off the equilibrium path.

Thus if a default occurs before the critical stage defined in the next chapter, it is off the equilibrium path, because in equilibrium both the bad bank and (obviously) the good one play safe with probability one in any stage before the critical one. Then the choice of putting $\rho_{n+1} = 1$, does not come from the Bayes' rule but is a free choice within the coherency required by the perfect Bayes-Nash rules.

Another information set off the equilibrium path is reached if after a default (on or off the equilibrium path), one or possibly more savers of the late generations play D without finding new defaults.

Here the choice to treat the bank as surely bad could be questionable: indeed suppose that the first (and unique) default happened before the critical stage and that a lot of savers after the critical stage played D without incurring a new default. The probability of such an event may be very small, thus this new evidence should perhaps induce us to reconsider the previous (free but unjustified!) choice of putting $\rho_{n+1} = 1$. This is just one example of the complications that may emerge when looking at beliefs out of equilibrium in multistage games.

Section 10

But let us look now at a possible account of the equilibrium path of the game: suppose that at the beginning ρ_1, the initial public information probability that the bank is bad, is lesser than the minimum value β of the boundary function; the saver deposits and both types of banks play sure: there is no default, but this conveys no new information to the saver of the second generation, whose beliefs do not update; he deposits too and both banks go on playing safe and so on until the game enters the tail. It is clear from 3b) and 9.2 that the saver of the first generation in the tail goes on depositing, while now the two types of banks separate: the good one plays safe, while the bad one switches to the risky strategy; this is the pattern until a default occurs: in this case the next generations of savers, coherently with 4b), recognize that the bank is surely of the bad type so they do not play and the game definitively ends.

Suppose on the contrary that there is some **(critical) stage** n in the body of the game such that $\beta z_{n+1} < \rho_1 < \beta z_n$ (see Fig. 11) It is clear that in any stage before n all savers played and both types of banks pooled at the safe strategy, so that $\rho_n = \rho_1$. Since then $\rho_n < \beta z_n$, the $n-th$ saver deposits too, but the bad bank now randomizes in such a way that the application of the Bayes rule 9.2 in the case of solvency, moves the beliefs to $\rho_{n+1} = \beta z_{n+1}$.

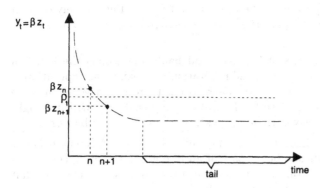

Fig. 11.

We could check that this is obtained simply by having recourse to the mixed strategy ϕ_n defined in 3a). We saw that in this case the next saver randomizes too: if the outcome of the randomization is to play N, the game definitively ends because the later savers do not want to play; if the outcome is D a new randomization of the bad bank occurs, and a chain of randomizations take place until the game reaches the tail or has an early end due to a default of the bad bank or to an outcome N of the saver's decision. In the tail the behavior is the one already discussed.

Undoubtedly someone will think that there is too much rationality and little intuition in "this complicated dance of mixed strategies" (Kreps (1992 pag. 541).

Anyway, here we offer a verbal explanation of the rationality behind the equilibrium strategies: the bad bank (as well as the good one) hopes to play as long as possible so as to arrive at the tail of the game, even at the cost of lowering some of the gains in the body.

To reach this goal it is not enough to play safe: it is necessary that the savers play D, and this is done only if low enough values of their beliefs that the bank is bad prevail. Moreover as we will immediately see looking at the boundary function, as time goes these beliefs should become smaller and smaller.

Then, until there is a low probability to be perceived as bad, the bank does not find it convenient to risk its reputation and plays safe (the advantage of playing risky in the single stage does not offset the tremendous loss of reputation suffered in case of default).

On the contrary if, initially or during the game, the beliefs that a bank is a bad bank is too high, then that bank has an incentive that the savers think it is going to incur some risk, as this is the only way a lucky bank may recover reputation pushing down the beliefs.

Of course the bank does not want to gain too great a reputation (at the cost of a big risk), but just what is useful: hence randomization rather than a pure risky strategy.

In turn the saver, by randomizing, keeps the bank under the threat that the game may end suddenly even if no default has been registered, thereby giving an incentive to the bad bank to incur the right default risk and not merely to bluff on this point by claiming to play risky and indeed playing safe (thus gaining undeserved reputation).

Summing up, if the beliefs of being a good bank (one minus the belief of being a bad bank) are a measure of the reputation of a bank, the reputation is **preserved** by playing safe, but is **increased** (and there is a need to increase reputation as we are coming close to the tail of the game) only with the right size of risky behavior (jointly with good luck!) on the part of the bad bank.

Note finally that the single randomizing saver is per se indifferent between depositing or not; nevertheless being part of an intergenerational agreement he plays in the interest of the community and does to the next savers exactly what he wishes the previous savers have done to him!

Section 11

A fine distinction between building and defending one's reputation is the core of a more complex and perhaps realistic model of a credit market game (Diamond 1989), an understanding of which is greatly enhanced by our previous analysis of the credit game.

In the original version the model explores the relations between borrowers intended as entrepreneurs and lenders intended as savers, but we find it convenient here to force the interpretation in which borrowers are intermediaries and lenders are savers.

There is now a cohort of three types of intermediaries: "good" with the opportunity to invest at any stage in safe loans with a gross return of s greater than the "reservation return" r^* (see below), "bad" with the opportunity to choose risky loans with a high return of $b > s$ or a low return (say 0), and expectation m lesser than r^*, and "Janus" which can choose between safe and risky projects.

There is a sequence of savers (potential lenders) who have the opportunity to gain elsewhere (not entering in the market) a reservation rate r^* (constant over time), and whose beliefs are at the beginning of the game based on public information on the relative frequencies of the three types, and later updated coherently with the prescriptions of the sequential equilibrium concept. At no stage will the savers accept lending at an expected rate (based on their beliefs) which is lesser than the reservation.

The game is played as follows:

1) at stage $h(h = 1, \ldots N)$ borrowers of type t offer a debt contract with face value r_{ht}; note that debt contracts are neither constrained to be "stationary" over time nor invariant with type;
2) the lenders choose which contracts to accept (which loans to finance) on the basis of their beliefs and contractual conditions;
3) the borrowers that have been financed (received money)select portfolios according to their opportunities; if they are of the Janus type they choose between safe and risky loans;
4) nature determines the outcome of risky loans; neither the choice nor the outcome of the loans are observable by lenders except indirectly in case of a default;
5) the borrowers observe the outcome of their loans and pay the debt, whenever possible, according to the contractual conditions; if the outcome of a risky loan is bad they are forced to default;
6) a new stage starts, with non defaulters offering new debt contracts and savers of the new generation updating beliefs on the basis of the information obtained (defaults in the previous stages and contracts offered in the current stage) and so on.

The savers' goal is to maximize the expected value of their single stage decision (they are active only for one stage),while the intermediary tries to maximize the present value of the expectation of profits over a finite (but supposedly long) horizon.

A sequential equilibrium supported by beliefs about the type of borrowers that are not a function of the rate offered (no signalling effect), reveals the following interesting features.

Everything is driven by the intuitive idea that the higher the cost of money – today and in the future – the higher the pressure for Janus borrowers to play risky, and that only borrowers perceived to be good or willing to play good with some probability are financed.

Coherently with the no signalling hypothesis, at any time all non defaulted borrowers pool and offer the same contract, based on the lower rate compatible with the beliefs of the savers. This means that all (indistinguishable) types of borrowers are financed.

After that the features of the equilibrium depend on the degree of adverse selection (percentage of bad players in the economy).

Precisely in the absence of adverse selection, or with negligible adverse selection, the qualitative features of the solution (we do not go here in numerical details) are very close to the ones of the credit game treated in the previous chapter: safe loans are chosen in the beginning (by a Janus bank), then there is a mid term period of mixed strategies before switching in the tail to the pure risky strategy. Defaulters if any are immediately fired off.

On the contrary, with widespread adverse selection, in the early stages the contractual return would be relatively high and this fact incentivates the Janus borrowers to play risky. As time goes by several of the bad and of the free-risky borrowers default.

Defaulted borrowers are definitively fired off. Thus the contractual rate falls and this makes it convenient for the Janus borrowers to play safe at least until the tail of the game where a new switch to risky loans takes place. This way, the game really goes back to mimic the behavior of the credit game already described.

Indeed, note that even if the Janus players use pure safe strategies, a mixed strategy effect derives from the existence of some bad players so that someone playing risky loans may be found at any stage at least until all bad players have defaulted.And in the mid term stages, that is enough to enforce the reputation of the Janus players which have survived the selection of the first period.

Summarizing, in the more interesting case of widespread adverse selection the initial goal is to play risky to exploit high rates. Survivors gain reputation and this in turn lowers rates of return. When this stage has been reached, after some periods of a good track record, reputation becomes a valuable asset.

Janus intermediaries now equipped with a good reputation do not want to risk loosing it and will play safe to defend this reputation. At the same time the reputation is freely (without risk) enforced by the behavior of the surviving bad banks.

The model developed by Diamond points out that a young firm does not really worry about reputation at first; some incentive to moral hazard may exist at the beginning; after some period the firm acquires a reputation that becomes a valuable immaterial asset which is defended by choosing safe projects.

A need to careful monitoring and control of the activities of financial intermediaries in the first years of their life emerges from this picture. Incidentally this is just what has been traditionally done in the regulation of insurance companies: at the beginning they are for example compelled to high reinsurance quotas which fall after say a period of five or ten years.

Section 12

The model of Diamond is a big step toward a formalized general theory of financial intermediaries in a game theoretic setting characterized by incomplete information, multi-stage non cooperative games in extensive form.

Here we offer a tentative picture of such a general approach.

1) There are intermediaries of various possible types (a concept here intended in a fuzzy sense): type embodies rates of returns of assets in the portfolio of the intermediary (a different ability in selecting assets implies different opportunity sets) and/or a different propensity to be fair (averse to opportunistic behavior), perhaps because of a different structure of payoffs (of the utility function). The percentage of the various types is public information.

2) Each intermediary offers savers of the first generation a menu of possible contracts, specifying for each one such things as:the (contingent to solvency) face value of the liabilities that he accepts, the guarantee fund that he wants to keep in order to grant payment of the liabilities, the plan of investments of the resources under its control, the price for any type and amount of liability.

3) Savers of the first generation, after having updated (though not necessarily changed) the probability distribution on the type of the intermediary on the basis of the signal deriving from the contractual menu (if meaningful), sign a contract (or none) and pay the contractual price.

4) The intermediary selects the assets, consintently or not with the agreed strategy (usually the strategies of the intermediary cannot be monitored by the saver).

5) Nature determines the global value of the intermediaries' at any stage assets (which is usually considered at least partially private information too).

6) The intermediary decides to honour its liabilities or otherwise defaults (bankrupcy penalties are levied in this case so that there is no incentive for the intermediary to be insolvent unless it is really short of resources). In case of default the intermediary is definitively off the market; otherwise a second stage is open and the intermediary offers savers of the second generation a menu of contracts, possibly contingent to the first stage result.

7) The savers of the second generation update their beliefs on the type of intermediary on the basis of the information obtained following the updating of the first stage.

8) The intermediary selects the second stage portfolio of assets....and so on.

It is remarkable that generations of savers do not overlap: any saver chooses a contract from the menu of his stage with the goal to maximize the expected utility of this stage; on the contrary the intermediary's goal is to maximize the present value of the expected utility of its strategy (the intermediary is not myopic).

Undoubtedly, the framework discussed is quite difficult to analyze in a formal quantitative setting; yet it omits some far from minor factors playing a decisive role in the real life of financial intermediaries. We sketch here only one such factor: generations of savers overlap.

Many intermediaries but especially life insurance and pension funds sign long term contracts whose maturity comes well after new contracts with new generations of customers have been signed; disclosure of partial results on the early generations may have a big influence on the updating of the intermediaries' characteristics by following generations, yet only a few of the first generation liabilities may have been paid in the meantime and maybe some shorter contracts of the young generations could end before the longest contracts of the old generations. Mismatching of overlapping generations should be carefully avoided. Yet it currently happens that the early generations enjoy a sort of "first in benefit", as a good provisional treatment without paying any cash outflow, seems to be quite efficient in capturing new customers of late generations.

Even more opportunistic moves have been found in open end real estate funds with new entrants paying not a fair value of their future claims, but the value of those earlier entrants, retiring timely after a largely ungrounded boom of their shares. Only a careful reflexion on reputation incentives may found efficient yet not too binding regulation rules for intermediaries with overlapping generation problems and more generally long lasting liabilities.

Acknowledgements. I wish to thank G. Forestieri and F. Ortu for helpful suggestions and comments, A. Cruccu and M. Florean, scholars of my game theory seminars, for valuable help in screening and reviewing related literature, Loredana Gaviglio for precious help with graphics and J. Douthwaite and A.P. De Luca for kind assistence in improving the English version of the paper. The usual warnings apply.
Financial support from MURST Fondi 40% 93-2-09-02-002 (Modelli di struttura per scadenza dei tassi di interesse) e 93-2-09-02-003 (Il rischio di tasso nell'assicurazione di portafoglio) and 60% (Rischio di tasso e rischio di insolvenza nella valutazione degli intermediari finanziari) is gratefully acknowledged.

References

Akerlof, G. (1970) 'The market for "lemons"', *Quarterly Journal of Economics*, 90:629-650.

Black, F., and Scholes, M. (1973) 'The pricing of options and corporate liabilities'. Journal of Political Economy, 81:637-659.

Ciocca, P. et al, (1991), 'Gruppo polifunzionale o Banca universale', Note Economiche 21:186-210.

Cho, I.-K. and Kreps, D. (1987) 'Signalling games and stable equilibria', *Quarterly Journal of Economics*, 102:179-222.

Diamond, D. (1984), 'Financial intermediation and delegated monitoring', Review of Economic Studies, 51:393-414.

Diamond, D. (1989), 'Reputation Acquisition in debt markets', Journal of Political Economy, 97:829-862.

Forestieri, G., 'Sistema finanziario e criteri allocativi:effetti sul grado di innovazione delle strutture produttive', (1993). Preprint Istituto di Economia degli intermediari finanziari. Université Bocconi Milano.

Fudenberg, D., and Tirole, J. 1991) 'Perfect Bayesian equilibrium and sequential equilibrium', *Journal of Economic Theory*, 53:236-260.

Fulghieri, P., and Rovelli, R., (1993), 'Le banche e la teoria dell'intermediazione', Preprint Istituto di Economia degli Intermediari Finanziari Universit2 Bocconi, Milano

Gibbons, R. (1992) Game Theory for Applied Economics.Princeton University Press, Princeton.

Greenwald, B., and Stiglitz, J.E. (1989), 'Information, Finance and Markets: The Architecture of Allocative Mechanisms'. Paper presented at the International conference on the history of enterprise, Terni, Italy.

Harrison, M., and Kreps, D. (1979), 'Martingales and arbitrage in multiperiod security markets', Journal of Economic Theory, 20:381-408.

Harrison, M., and Pliska, S. (1981), 'Martingales and stochastic integrals in the theory of continuous trading', Stochastic Processes and their applications, 11:215-260.

Kreps, D., (1992): A course in microeconomic theory. Harvester Wheatsheaf New York.

Kreps, and Wilson, R. (1982) 'Sequential equilibrium', *Econometrica*, 50:863-894.

Leland, H., and Pyle, D. (1977), 'Informational asymmetries, financial structure and financial intermediation', *Journal of Finance*, 32:371-387.

Lucas, J.D., and McDonald, R., (1992), 'Bank financing and investment decisions with asymmetric information about loan quality', Rand Journal of Economics, 23:86-105.

Masera, R., (1991), 'Innovazione finanziaria e ruolo delle banche nella finanza d'impresa', Note Economiche Monte dei Paschi, 21:173-185.

Myers, S., and Majluf, N., (1984), 'Corporate finance and investment decisions when firms have informations that investors do not have', Journal of Financial Economics, 13:187-221.

Markovitz, H. (1952) 'Portfolio selection', Journal of Finance 7:77-91.
Merton, R. (1973)'Theory of rational option pricing', Bell Journal of Economics and Management Science 4:141-183.

Ortu, F., and Pressacco, F., (1990) 'Ammissibilità e completezza dei mercati finanziari nel prezzamento delle opzioni'. Scritti in omaggio a Luciano Daboni Lint, Trieste.

Parigi, B.M., (1992) 'Repeated lending with limited liability under imperfect monitoring', Economic Notes, 21:468-489.

Ross, S., (1976) 'The arbitrage theory of capital asset pricing', Journal of Economic Theory 17:341-360.

Sabani, L. (1991), 'Financial intermediation and moral hazard: the role of market forces in implementing self enforcing contracts between bank and depositors', Economic Notes, 20:497-523.

Sharpe, W. (1964) 'Capital asset prices:a theory of market equilibrium under uncertainty', Journal of Finance 19:425-442.

Sharpe, W. (1991) 'Assessing the style and performance of U.S. mutual funds with an asset class factor model', Paper presented at the 2nd international conference I.M.I. Group, Rome.

Spence, A.M. (1973) 'Job market signalling', *Quarterly Journal of Economics*, 87:355-374.

Terlizzese, D. (1988), 'Delegated Screening and reputation in a theory of financial intermediaries', Temi di discussione Banca d'Italia 111.

Van Damme, E., (1991) Stability and Perfection of Nash equilibria. Springer Verlag. Berlin.

Printing: Weihert-Druck GmbH, Darmstadt
Binding: Buchbinderei Schäffer, Grünstadt